Frederick Chapman

On some Foraminifera obtained by the Royal Indian Marine

Survey's S. S.

Frederick Chapman

On some Foraminifera obtained by the Royal Indian Marine Survey's S. S.

ISBN/EAN: 9783741194450

Manufactured in Europe, USA, Canada, Australia, Japa

Cover: Foto ©berggeist007 / pixelio.de

Manufactured and distributed by brebook publishing software
(www.brebook.com)

Frederick Chapman

On some Foraminifera obtained by the Royal Indian Marine

Survey's S. S.

two birds being sexes of the same species.' I may add that the present specimen is the only one of this hybrid that I have ever handled, or, to the best of my recollection, ever seen."

The Rev. T. R. R. Stebbing exhibited a specimen of a species of *Peripatus* from Antigua.

The following papers were read :—

1. On some Foraminifera obtained by the Royal Indian Marine Survey's S.S. 'Investigator,' from the Arabian Sea, near the Laccadive Islands. By FREDERICK CHAPMAN, F.R.M.S.

[Received November 7, 1894.]

(Plate I.)

On the 25th of July, 1893, I received some samples of deep-sea soundings from Mr. T. H. Holland, F.G.S., Assistant-Superintendent of the Geological Survey of India, who, in conjunction with Dr. Alcock, Surgeon-Naturalist in the Royal Indian Marine Survey Department, has very kindly placed the material in my hands for description.

The results of a somewhat exhaustive examination of the soundings appear to be of sufficient interest for publication. Moreover, the locality from whence these soundings were obtained is sufficiently out of the path of former expeditions to make the list useful. The soundings were obtained by the Royal Indian Marine Survey's steamship 'Investigator' from a limited area near the Lakadivh (Laccadive) Islands, viz., 15° 30′ 4″ to 8° 21′ 3″ N. lat. and 75° 42′ 5″ to 71° 09′ 3″ E. long.

The depths at which the soundings were obtained have not been recorded, but they did not exceed 1238 fathoms. This absence of the record of depths is explained by the fact that the material was originally sent to Mr. Holland for his opinion regarding the supposed occurrence of submarine volcanoes in that part of the Arabian Sea, the search for Foraminifera being undertaken subsequently. Samples of the soundings were sent to me after having been washed for the purpose just stated.

Mr. Holland has also kindly furnished me with the following notes concerning the temperature of the area from whence the material was taken.—"The lowest bottom temperature recorded was 37° F., several times at about 1130 fathoms; the surface temperature being about 78°–80° F."

The material was sent in six bottles, each sample appearing to represent a mixture of several distinct soundings. The samples are labelled thus :—

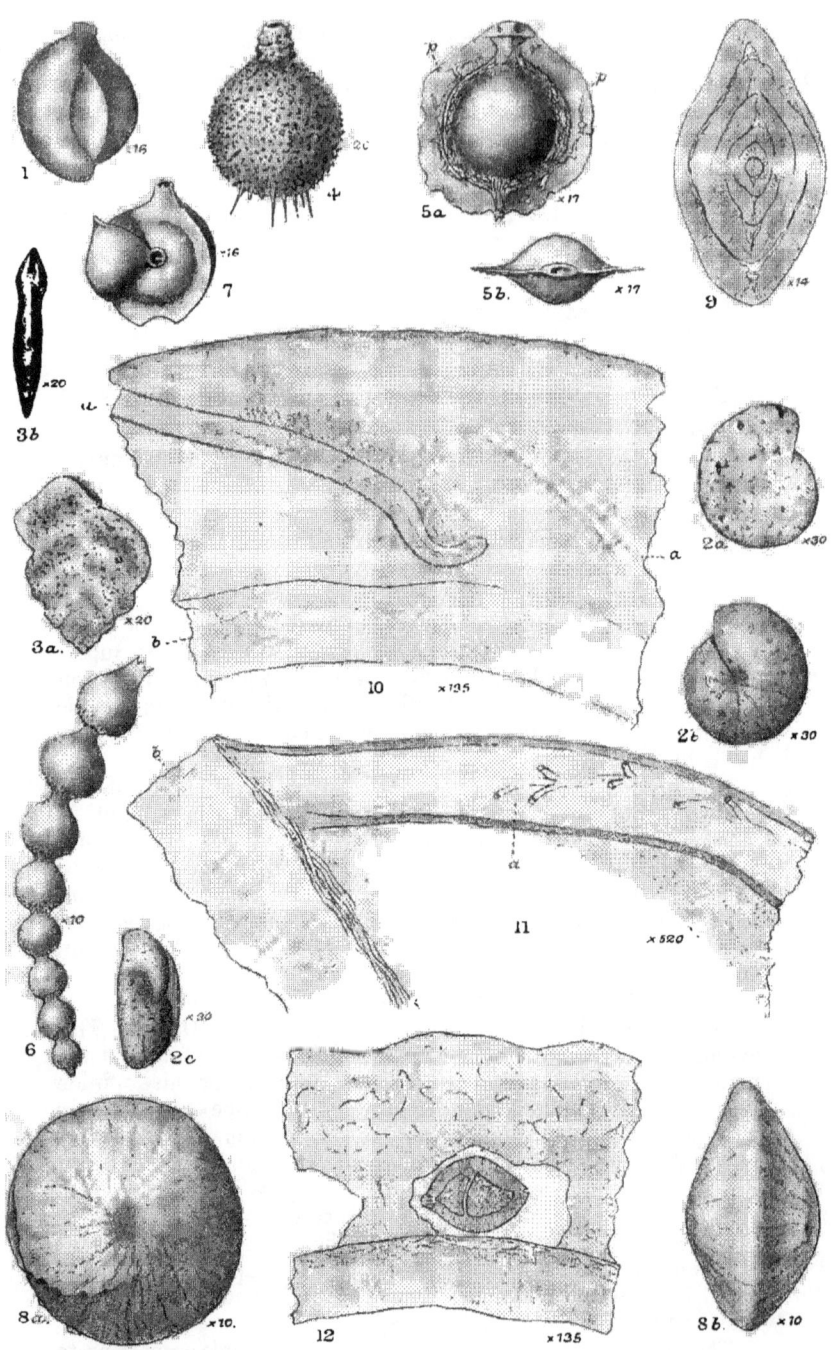

F.Chapman del.ad nat.
E.C Knight lith.

West.Newman imp.

Foraminifera from the Arabian Sea.

Sample No. 1.

Nos. 1–4, 6–9, 10, 14, 15–18, 19, 20, 22, 26, 49, 50–54, 58, 59.
" *Globigerina* ooze.   Terrigenous deposit.   *Globigerina* ooze.
Green mud.   *Globigerina* ooze."

Sample No. 2.

Nos. 11, 23, 55.
"Green mud.   Broken coral.   S. crl. (= (?) Sandy mud with coral).   Broken coral."

Sample No. 3.

Nos. 23–25, 28–30, 62.
" Green ooze."

Sample No. 4.

Nos. 27, 32, 34, 35, 38–44, 57, 60.
" Grey ooze."

Sample No. 5.

Nos. 31, 45–48.
" Brown ooze."

Sample No. 6.

Nos. 47–53.
" Brown ooze.   *Globigerina* ooze."

The washed soundings, as received, consisted of, in each case, a nearly pure assemblage of foraminiferal shells ; with the exception of Sample No. 2, which was composed mainly of coral detritus with some Foraminifera.

The other organisms noticed in the material during the search for Foraminifera were the following.—

Sample No. 1.

Ostracoda.—*Pontocypris* (?) *subreniformis*, G. S. Brady.
   Macrocypris tenuicauda, G. S. B.
   Bairdia attenuata, G. S. B.
   —— *hirsuta*, G. S. B.
   Cytherella polita, G. S. B.
Also Radiolaria of 3 species.

Sample No. 3.

Ostracoda.—*Cythere dictyon*, G. S. B.

Sample No. 4.

Some fish otoliths.
Pteropoda.—*Clio (Styliola) subula* (Quoy & Gaimard) and *Cavolinia uncinata* (Rang).
Ostracoda.—*Bairdia hirsuta*, G. S. B.
   Cythere dictyon, G. S. B.
   —— *acanthoderma*, G. S. B.
   —— (?) *serrulata*, G. S. B.
   Cytherella polita, G. S. B.

Sample No. 5.

Ostracoda.—*Pontocypris faba* ? (Reuss).
Cythere (?) *serratula*, G. S. B.
Krithe hyalina, G. S. B.
Xestoleberis expansa, G. S. B.
Pseudocythere caudata, G. O. Sars.
Also Radiolaria of 6 species.

Sample No. 6.

Ostracoda.—*Pontocypris faba* ? (Reuss).
Macrocypris decora, G. S. B.
Bairdia foveolata, G. S. B.
—— *minima*, G. S. B.
—— *hirsuta*, G. S. B.
Cythere (?) *serratula*, G. S. B.
Cythere dictyon, G. S. B.
—— *radula*, G. S. B.
Krithe hyalina, G. S. B.
—— *producta*, G. S. B.
Cytherella polita, G. S. B.

A striking feature of the foraminiferal fauna of the Arabian Sea is the presence therein of a number of forms which have already been described by Dr. Conrad Schwager[1] from marine clay-beds, of late Pliocene age, on the northern coast of Kar Nicobar[2].

Although many of the species recorded from those fossil beds were subsequently found by Dr. Brady in the 'Challenger' soundings, yet there still remained at least five well-defined forms recorded by Dr. Schwager from the above-mentioned Pliocene beds, and these are noted here for the first time as recent Foraminifera.

There are altogether eight species new to the list of recent Foraminifera, but which have formerly been recorded as fossils: these are:—*Textularia lythostrotum* (Schwager), *Cassidulina murrhina* (Schwager), *Lagena capillosa* (Schwager), *Nodosaria (Dentalina) adolphina* (d'Orb.), *Nodosaria ovulata*, Sherb. & Chap., *Nodosaria (D.) acicula*, Lam., *Polymorphina fusiformis* (Römer), and *Calcarina nicobarensis*, Schwager.

Besides these there are two new species and three new varieties.

A list of Foraminifera obtained from the Bay of Bengal by H.M.S. 'Investigator' (from lat. 17° 34' N., long. 87° 59' E.) is given by Dr. John Murray in the 'Scottish Geographical Magazine' for August 1889. The material examined was a blue mud, obtained at a depth of 1300 fathoms. Amongst the thirty-seven species therein enumerated there are, however, no Tertiary species such as were obtained from the Arabian-Sea deposits.

In concluding these preliminary remarks it only remains for

[1] "Fossile Foraminiferen von Kar-Nicobar." Novara-Exped., geol. Theil, vol. ii. 1866, p. 187. 4to, Wien.
[2] F. von Hochstetter, *op. cit.* p. 88.

me to express my deep obligations to Professor T. Rupert Jones, F.R.S., for his kind help and advice during the writing of this paper.

In the following detailed account of the species, references are made chiefly to the monograph of Dr. H. B. Brady, since it is a standard and easily accessible work.

FORAMINIFERA.

NUBECULARIA, Defrance.

1. NUBECULARIA LUCIFUGA, Defrance.

Nubecularia lucifuga, Brady, 1884, Chall. Rep. vol. ix. p. 134, pl. i. figs. 9–16.

A single specimen of this species occurred in Sample No. 1.

BILOCULINA, d'Orbigny.

2. BILOCULINA DEPRESSA, d'Orbigny.

Biloculina depressa, Brady, 1884, Chall. Rep. vol. ix. p. 145, pl. ii. figs. 12, 15–17, pl. iii. figs. 1, 2.

This species attains a great size in these deposits, many of the specimens being as large as $\frac{1}{10}$ inch in diameter, measured across the face. Common in Sample No. 1 ; frequent, No. 3 ; common, No. 4 ; rare, No. 5 ; frequent, No. 6.

3. BILOCULINA DEPRESSA, d'Orb., var. MURRHYNA, Schwager.

Biloculina murrhyna, Schwager, 1866, Novara-Exped., geol. Theil, vol. ii. p. 203, pl. iv. fig. 15 *a–c*.

B. depressa, var. *murrhyna*, Brady, 1884, Chall. Rep. vol. ix. p. 146, pl. ii. figs. 10, 11 ; Schlumberger, 1885, Bull. Soc. Géol. France, sér. 3, vol. xiii. p. 238, figs. 9, 10, p. 290, figs. 15, 16 ; Schlumberger, 1891, Mém. Soc. Zool. France, vol. iv. p. 552, pl. ix. figs. 52–54.

This deep-water form has been recorded from the Atlantic and Pacific Oceans from depths between 1180 and 1900 fathoms (Brady). It was also found fossil in the Island of Kar Nicobar (Schwager).

Frequent in Sample No. 1 ; frequent, No. 4 ; rare, No. 6.

4. BILOCULINA DEPRESSA, d'Orb., var. SERRATA, Brady.

Biloculina depressa, var. *serrata*, Brady, 1884, Chall. Rep. vol. ix. p. 146, pl. iii. fig. 3 *a–c*.

This, another deep-water form, is rare in these deposits in Sample No. 1 ; very rare, No. 3 ; rare, No. 6.

5. BILOCULINA TUBULOSA, Costa.   (Plate I. figs. 1, 7.)

Biloculina tubulosa, Brady, 1884, Chall. Rep. vol. ix. p. 147, pl. iii. fig. 6 *a–c*.

This species has before been found off Kandavu, Fiji Islands, at 210 fathoms (Brady).

An abnormal variety of this species was found in Sample No. 1 (see Pl. I. fig. 7), in which the last two chambers have been twisted; and both possess characteristic tubular apertures, so that the specimen has the appearance of two forms intergrown at right angles to one another.

This species is frequent in Sample No. 1 ; rare, No. 3 ; common, No. 4 ; very rare, No. 5 ; frequent, No. 6.

6. BILOCULINA RINGENS (Lamarck).

Biloculina ringens, Brady, 1884, Chall. Rep. vol. ix. p. 142, pl. ii. figs. 7, 8.

This form occurs very rarely in Sample No. 6.

7. BILOCULINA RINGENS (Lam.), var. STRIOLATA, Brady.

Biloculina ringens, var. *striolata*, Brady, 1884, Chall. Rep. vol. ix. p. 143, pl. iii. figs. 7, 8.

Previously found in the Pacific from depths of from 6 to 8 fathoms (Brady).

Very rare in Sample No. 3; very rare, No. 4; rare, No. 6.

8. BILOCULINA COMATA, Brady.

Biloculina comata, Brady, 1884, Chall. Rep. vol. ix. p. 144, pl. iii. fig. 9 a, b.

This species occurs very rarely in Sample No. 6.

SPIROLOCULINA, d'Orbigny.

9. SPIROLOCULINA ROBUSTA, Brady.

Spiroloculina robusta, Brady, 1884, Chall. Rep. vol. ix. p. 150. pl. ix. figs. 7, 8.

This species was described by Dr. Brady from specimens found off Culebra Island, West Indies, 390 fathoms.

Typical examples of *S. robusta* were frequent in Sample No. 1.

10. SPIROLOCULINA ANTILLARUM, d'Orbigny.

Spiroloculina antillarum, Brady, 1884, Chall. Rep. vol. ix. p. 155, pl. x. fig. 21 a, b.

This species was found in Sample No. 1, rare.

11. SPIROLOCULINA LIMBATA, d'Orbigny.

Spiroloculina limbata, Brady, 1884, Chall. Rep. vol. ix. p. 150, pl. ix. figs. 15–17.

A specimen of the ordinary typical form was found in Sample No. 1.

12. SPIROLOCULINA GRATA, Terquem.

Spiroloculina grata, Brady, 1884, Chall. Rep. vol. ix. p. 155, pl. x. figs. 16, 17, 22, 23.

This species generally affects areas round coral-reefs, and is

recorded from the shallow waters of the Red Sea, from shore-sand on the E. coast of Madagascar, and in various parts of the Pacific Ocean (Brady).
S. grata occurred in Sample No. 1, very rare ; No. 2, frequent.

13. SPIROLOCULINA ARENARIA, Brady.

Spiroloculina arenaria, Brady, 1884, Chall. Rep. vol. ix. p. 153, pl. viii. fig. 12.

This moderately-shallow-water form has been noted from the Fiji Islands, Raine Island, and the Philippine Islands (Brady).
It occurred in Sample No. 4, rare.

14. SPIROLOCULINA ASPERULA, Karrer.

Spiroloculina asperula, Brady, 1884, Chall. Rep. vol. ix. p. 152, pl. viii. figs. 13, 14, and 11 ?
Found in Sample No. 4, rare.

MILIOLINA, Williamson.

15. MILIOLINA TRIGONULA (Lam.).

Miliolina trigonula, Brady, 1884, Chall. Rep. vol. ix. p. 164, pl. iii. figs. 14–16.
This form occurs in Sample No. 6, very rare.

16. MILIOLINA INSIGNIS, Brady.

Miliolina insignis, Brady, 1884, Chall. Rep. vol. ix. p. 165, pl. iv. figs. 8, 10.
Found in Sample No. 2. very rare.

17. MILIOLINA TRICARINATA (d'Orb.).

Miliolina tricarinata, Brady, 1884, Chall. Rep. vol. ix. p. 165, pl. iii. fig. 17 a, b.
Found in Sample No. 1, frequent; No. 2, very rare; No. 5, rare.

18. MILIOLINA CIRCULARIS (Bornemann).

Miliolina circularis, Brady, 1884, Chall. Rep. vol. ix. p. 160, pl. iv. fig. 3 a-c, pl. v. figs. 13, 14 ?
Found in Sample No. 1, very rare.

19. MILIOLINA AUBERIANA (d'Orb.).

Miliolina auberiana, Brady, 1884, Chall. Rep. vol. ix. p. 162, pl. v. figs. 8, 9.
Found in Sample No. 1, rare ; No. 4, rare ; No. 5, very rare.

20. MILIOLINA CUVIERIANA (d'Orb.).

Miliolina cuvieriana, Brady, 1884, Chall. Rep. vol. ix. p. 162, pl. v. fig. 12 a-c.
Found in Sample No. 5, very rare ; No. 6, very rare.

21. MILIOLINA VENUSTA (Karrer).
Miliolina venusta, Brady, 1884, Chall. Rep. vol. ix. p. 162, pl. v. figs. 5, 7.
Found in Sample No. 1, rare.

22. MILIOLINA GRACILIS (d'Orb.).
Miliolina gracilis, Brady, 1884, Chall. Rep. vol. ix. p. 160, pl. v. fig. 3 *a–c*.
Found in Sample No. 1, very rare.

23. MILIOLINA AMYGDALOIDES, Brady.
Miliolina amygdaloides, Brady, 1884, Chall. Rep. vol. ix. p. 163, pl. vi. fig. 10 *a, b*.
Found in Sample No. 1, very rare.

24. MILIOLINA BICORNIS (W. & J.).
Miliolina bicornis, Brady, 1884, Chall. Rep. vol. ix. p. 171, pl. v. figs. 9, 11, 12.
Found in Sample No. 1, very rare.

25. MILIOLINA SCHREIBERSIANA (d'Orb.).
Triloculina schreibersiana, d'Orbigny, 1839, Foram. Cuba, p. 174, pl. ix. figs. 20–22.
This species was described by d'Orbigny from specimens obtained out of the shore-sand from the Island of Cuba. It was not met with in any of the soundings obtained by the 'Challenger.' As might be expected, this species occurs in the shallow-water deposits of Sample No. 2 from the Arabian Sea, and is very rare.

26. MILIOLINA UNDOSA (Karrer).
Miliolina undosa, Brady, 1884, Chall. Rep. vol. ix. p. 176, pl. vi. figs. 6–8.
Found in Sample No. 1, frequent.

27. MILIOLINA LINNÆANA (d'Orb.).
Miliolina linnæana, Brady, 1884, Chall. Rep. vol. ix. p. 174, pl. vi. figs. 15–20.
This shallow-water form has been before noted from the coral islands of the Pacific, &c. (Brady). In the gatherings from the Arabian Sea it occurs in Sample No. 1, and is very rare.

28. MILIOLINA RETICULATA (d'Orb.).
Miliolina reticulata, Brady, 1884, Chall. Rep. vol. ix. p. 177, pl. ix. figs. 2–4.
This species is essentially a shallow-water form, being found in shore-sands and in the neighbourhood of coral-reefs.
Found in Sample No. 1, frequent.

29. MILIOLINA PARKERI, Brady.

Miliolina parkeri, Brady, 1884, Chall. Rep. vol. ix. p. 177, pl. vii. fig. 14.
This form is also usually associated with coral-reefs.
Found in Sample No. 1, very rare.

30. MILIOLINA RUPERTIANA, Brady.

Miliolina rupertiana, Brady, 1884, Chall. Rep. vol. ix. p. 178, pl. vii. figs. 7-12.
Found in Sample No. 1, very rare.

OPHTHALMIDIUM, Kubler.

31. OPHTHALMIDIUM INCONSTANS, Brady.

Ophthalmidium inconstans, Brady, 1884, Chall. Rep. vol. ix. p. 189, pl. xii. figs. 5, 7, 8.
Found in Sample No. 1, very rare.

SIGMOÏLINA, Schlumberger.

32. SIGMOÏLINA SIGMOIDEA (Brady).

Planispirina sigmoidea, Brady, 1884, Chall. Rep. vol. ix. p. 197, pl. ii. figs. 1-3; woodcut fig. 5 c.
Sigmoïlina (*Planispirina*) *sigmoidea*, Schlumberger, 1887, Bull. Soc. Zool. France, vol. xii. p. 118, pl. vii. figs. 9-11; woodcuts, figs. 1-5.
Found in Sample No. 1, very rare.

33. SIGMOÏLINA CELATA (Costa).

Spiroloculina celata, Costa, 1855, Mem. Accad. Napoli, vol. ii. p. 126, pl. i. fig. 14; 1856, Atti dell' Accad. Pont. vol. vii. pl. xxvi. fig. 5.
Quinqueloculina asperula et *rugosa*, Schwager, 1866, Novara-Exped., geol. Theil, vol. ii. pp. 203, 266, pl. iv. fig. 16 a-c.
Planispirina celata, Brady, 1884, Chall. Rep. vol. ix. p. 197, pl. viii. figs. 1-4.
Sigmoïlina (*Planispirina*) *celata*, Schlumberger, 1887, Bull. Soc. Zool. France, vol. xii. p. 481, pl. vii. figs. 12-14; woodcuts, figs. 6, 7.
This is another of the forms found fossil in the Island of Kar Nicobar by Dr. Schwager.
Found in Sample No. 1, common; No. 3, rare; No. 4, rare; No. 5, rare; No. 6, rare.

CORNUSPIRA, Schultze.

34. CORNUSPIRA CARINATA (Costa).

Cornuspira carinata, Brady, 1884, Chall. Rep. vol. ix. p. 201 pl. xi. fig. 4 a, b.
Found in Sample No. 1, very rare.

ORBITOLITES, Lamarck.

35. ORBITOLITES COMPLANATA, Lamarck.

Orbitolites complanata, Brady, 1884, Chall. Rep. vol. ix. p. 218, pl. xvi. figs. 1–6.

Found in Sample No. 1, common.

36. ORBITOLITES MARGINALIS (Lam.).

Orbitolites marginalis, Brady, 1884, Chall. Rep. vol. ix. p. 214, pl. xv. figs. 1–5.

Found in Sample No. 1, rare.

ALVEOLINA, d'Orbigny.

37. ALVEOLINA MELO (F. & M.).

Alveolina melo, Brady, 1884, Chall. Rep. vol. ix. p. 223, pl. xvii. figs. 13–15.

This species is usually found in coral-sands down to a depth of 40 fathoms (Brady).

Found in Sample No. 1, very rare.

38. ALVEOLINA BOSCII (Defr.).

Alveolina boscii, Brady, 1884, Chall. Rep. vol. ix. p. 222, pl. xvii. figs. 7–12.

This species also is associated with coral-sands.

Found in Sample No. 1, common : No. 2, rare.

TECHNITELLA, Norman.

39. TECHNITELLA MELO, Norman.

Technitella melo, Brady, 1884, Chall. Rep. vol. ix. p. 246, pl. xxv. fig. 7 *a*, *b*.

This species is rare, and has been recorded off Ascension Island at 420 fathoms (Brady); and from the N. Atlantic, S. of the Rockall Bank, at 1215 fathoms (Norman).

Found in Sample No. 4, very rare.

40. TECHNITELLA RAPHANUS, Brady.

Technitella raphanus, Brady, 1884, Chall. Rep. vol. ix. p. 247, pl. xxv. figs. 13, 14.

T. raphanus has been recorded from Kandavu, Fiji Ids., 210 fathoms (Brady).

In the soundings from the Arabian Sea it was found in Sample No. 4, frequent.

BATHYSIPHON, Sars.

41. BATHYSIPHON FILIFORMIS, Sars.

Bathysiphon filiformis, Brady, 1884, Chall. Rep. vol. ix. p. 248, pl. xxvi. figs. 15–20.

Found in Sample No. 4, very rare.

PSAMMOSPHÆRA, Schulze.

42. PSAMMOSPHÆRA FUSCA, F. E. Schulze.

Psammosphæra fusca, Brady, 1884, Chall. Rep. vol. ix. p. 249,
pl. xviii. figs. 1–8.
Found in Sample No. 1, very rare.

SACCAMMINA, M. Sars.

43. SACCAMMINA SPHÆRICA, M. Sars.

Saccammina sphærica, Brady, 1884, Chall. Rep. vol. ix. p. 253,
pl. xxiii. figs. 11–17.
Found in Sample No. 1, rare.

44. SACCAMMINA SOCIALIS, Brady.

Saccammina socialis, Brady, 1884, Chall. Rep. vol. ix. p. 255,
pl. xviii. figs. 18, 19.
This species has been hitherto found in the North Atlantic and
North Pacific Oceans (Brady).
It was found in Sample No. 5, very rare.

HYPERAMMINA, H. B. Brady.

45. HYPERAMMINA ELONGATA, Brady.

Hyperammina elongata, Brady, 1884, Chall. Rep. vol. ix. p. 257,
pl. xxiii. figs. 4, 7–10.
Found in Sample No. 1, rare; No. 4, common; No. 5, very
rare; No. 6, rare.

46. HYPERAMMINA RAMOSA, Brady.

Hyperammina ramosa, Brady, 1884, Chall. Rep. vol. ix. p. 261,
pl. xxiii. figs. 15–19.
Found in Sample No. 1, frequent; No. 3, rare; No. 4, very
common; No. 6, rare.

47. HYPERAMMINA ARBORESCENS (Norman).

Hyperammina arborescens, Brady, 1884, Chall. Rep. vol. ix.
p. 262, pl. xxviii. figs. 12, 13.
Found in Sample No. 1, rare; No. 4, very rare.

MARSIPELLA, Norman.

48. MARSIPELLA ELONGATA, Norman.

Marsipella elongata, Brady, 1884, Chall. Rep. vol. ix. p. 264,
pl. xxiv. figs. 10–19.
Found in Sample No. 4, rare.

RHABDAMMINA, M. Sars.

49. RHABDAMMINA DISCRETA, Brady.

Rhabdammina discreta, Brady, 1884, Chall. Rep. vol. ix. p. 268, pl. xxii. figs. 7–10.

Found in Sample No. 1, very rare; No. 4, frequent; No. 6, very rare.

RHIZAMMINA, H. B. Brady.

50. RHIZAMMINA INDIVISA, Brady.

Rhizammina indivisa, Brady, 1884, Chall. Rep. vol. ix. p. 277, pl. xxix. figs. 5–7.

Found in Sample No. 1, very rare; No. 4, common; No. 5, very rare.

REOPHAX, Montfort.

51. REOPHAX DIFFLUGIFORMIS, Brady.

Reophax difflugiformis, Brady, 1884, Chall. Rep. vol. ix. p. 289, pl. xxx. figs. 1–5.

The tests of these specimens from the Arabian Sea are composed of tiny embryonic globigerine shells in all cases excepting that from Sample No. 1, in which the specimen is composed of sandy and spicular material.

Found in Sample No. 1, very rare; No. 4, very rare; No. 5, very rare; No. 6, very rare.

52. REOPHAX SCORPIURUS (Montfort).

Reophax scorpiurus, Brady, 1884, Chall. Rep. vol. ix. p. 291, pl. xxx. figs. 12–17.

It is possible that one of the specimens referred to the above species, that from Sample No. 1, more properly belongs to *R. arctica*, Brady, since it has the usual tapering shell but strongly compressed. It differs, however, from the typical *R. arctica* in being quite as large as the well-known *R. scorpiurus*. The other specimens met with are of the ordinary type form.

Found in Sample No. 1, very rare; No. 4, frequent; No. 6, frequent.

53. REOPHAX SPICULIFERA, Brady.

Reophax spiculifera, Brady, 1884, Chall. Rep. vol. ix. p. 295, pl. xxxi. figs. 16, 17.

The usual form of this species has more or less cylindrical chambers; but some of the specimens from the Arabian Sea show a tendency to pass over into the form of *R. nodulosa*, Brady, whilst retaining the spicular test. It is therefore difficult to determine to which of the two above-mentioned species some of the specimens belong; the cylindrical form of the segments is here taken as the distinguishing character irrespective of the nature of the test.

R. spiculifera has hitherto been found off Kandavu, Fiji Islands, and Tahiti, Society Islands (Brady).
Found in Sample No. 1, rare; No. 4, very rare; No. 6, very rare.

54. REOPHAX DISTANS, Brady.

Reophax distans, Brady, 1884, Chall. Rep. vol. ix. p. 296, pl. xxxi. figs. 18–22.
Found in Sample No. 4, very rare.

55. REOPHAX NODULOSA, Brady.

Reophax nodulosa, Brady, 1884, Chall. Rep. vol. ix. p. 294, pl. xxxi. figs. 1–9.

As previously stated, I have included under this specific name some specimens which have the test formed mainly, if not entirely, of broken sponge-spicules but possessing oval or pyriform chambers. Associated with these are many specimens which have an arenaceous test, and which are therefore quite typical in character.
Found in Sample No. 1, frequent; No. 6, very rare.

56. REOPHAX DENTALINIFORMIS, Brady.

Reophax dentaliniformis, Brady, 1884, Chall. Rep. vol. ix. p. 293, pl. xxx. figs. 21, 22.
Found in Sample No. 1, rare; No. 3, very rare; No. 4, very rare; No. 5, very rare.

57. REOPHAX BACILLARIS, Brady.

Reophax bacillaris, Brady, 1884, Chall. Rep. vol. ix. p. 293, pl. xxx. figs. 23, 24.
Found in Sample No. 1, very rare.

58. REOPHAX PILULIFERA, Brady.

Reophax pilulifera, Brady, 1884, Chall. Rep. vol. ix. p. 292, pl. xxx. figs. 18–20.
Found in Sample No. 1, common.

HAPLOPHRAGMIUM, Reuss.

59. HAPLOPHRAGMIUM GLOMERATUM, Brady.

Haplophragmium glomeratum, Brady, 1884, Chall. Rep. vol. ix. p. 309, pl. xxxiv. figs. 15–18.
Found in Sample No. 6, rare.

60. HAPLOPHRAGMIUM LATIDORSATUM (Bornemann).

Haplophragmium latidorsatum, Brady, 1884, Chall. Rep. vol. ix. p. 307, pl. xxxiv. figs. 7–10, 14.
Found in Sample No. 1, rare; No. 4, common; No. 6, rare.

61. HAPLOPHRAGMIUM GLOBIGERINIFORME (P. & J.).

Haplophragmium globigeriniforme, Brady, 1884, Chall. Rep. vol.
ix. p. 312, pl. xxxv. figs. 10, 11.
This species has been recorded by Parker and Jones from the
Red Sea at 557 and 678 fathoms.
Found in Sample No. 1, rare; No. 6, rare.

62. HAPLOPHRAGMIUM CANARIENSE (d'Orb.).

Haplophragmium canariense, Brady, 1884, Chall. Rep. vol. ix.
p. 310, pl. xxxv. figs. 1–5.
Found in Sample No. 1. very rare: No. 4, very rare; No. 6,
very rare.

63. HAPLOPHRAGMIUM TURBINATUM, Brady.

Haplophragmium turbinatum, Brady, 1884, Chall. Rep. vol. ix.
p. 312, pl. xxxv. fig. 9 *a–c*.
Found in Sample No. 1. very rare : No. 4. very rare ; No. 6,
very rare.

64. HAPLOPHRAGMIUM ROTULATUM, Brady.

Haplophragmium rotulatum, Brady, 1884, Chall. Rep. vol. ix.
p. 306, pl. xxxiv. figs. 5, 6.
Found in Sample No. 1, very rare ; No. 4, very rare; No. 6,
very rare.

65. HAPLOPHRAGMIUM SCITULUM, Brady.

Haplophragmium scitulum, Brady, 1884, Chall. Rep. vol. ix.
p. 308, pl. xxxiv. figs. 11–13.
Found in Sample No. 1, rare ; No. 5, very rare ; No. 6, rare.

66. HAPLOPHRAGMIUM EMACIATUM, Brady.

Haplophragmium emaciatum, Brady, 1884, Chall. Rep. vol. ix.
p. 305, pl. xxxiii. figs. 26–28.
This species has been hitherto known only from the West Indies
(Brady).
Found in Sample No. 5, very rare.

67. HAPLOPHRAGMIUM AGGLUTINANS (d'Orb.).

Haplophragmium agglutinans, Brady, 1884, Chall. Rep. vol. ix.
p. 301, pl. xxxii. figs. 19–26.
Found in Sample No. 1, very rare.

68. HAPLOPHRAGMIUM TRUNCATULINIFORME, sp. nov.   (Plate I.
fig. 2 *a–c*.)

Test Rotaliform. Slightly concave on the superior, and strongly
convex on the inferior face ; the latter with a distinct umbilical
depression. Only the last convolution, which consists of twelve
chambers, can be seen on either face. Aperture strongly arched,
and confined almost entirely to the inferior face. Walls arenaceous,

of a yellow-brown colour, composed of fine material but with a few included coarser grains of a dark colour.   Edge of test rounded. Diameter $\frac{1}{42}$ inch ('6 mm.).

This species supplies a link in the chain of isomorphs of the hyaline and arenaceous groups of Foraminifera, since it bears the same relation to a typical *Truncatulina* or a *Rotalia* as *Haplophragmium globigeriniforme* does to *Globigerina*.

H. truncatuliniforme is represented in the Arabian-Sea soundings by only one example, from Sample No. 6.

PLACOPSILINA, d'Orbigny.

69. PLACOPSILINA CENOMANA, d'Orbigny.

Placopsilina cenomana, Brady, 1884, Chall. Rep. vol. ix. p. 315, pl. xxxvi. figs. 1–3.

Found in Sample No. 6, very rare.

THURAMMINA, Brady.

70. THURAMMINA PAPILLATA, Brady.

Thurammina papillata, Brady, 1884, Chall. Rep. vol. ix. p. 321, pl. xxxvi. figs. 7–18.

Found in Sample No. 4, rare.

HORMOSINA, Brady.

71. HORMOSINA CARPENTERI, Brady.

Hormosina carpenteri, Brady, 1884, Chall. Rep. vol. ix. p. 327, pl. xxxix. figs. 14–18.

It is interesting to record this species from the Arabian Sea, since it has previously been almost entirely confined to soundings from the North Atlantic.

Found in Sample No. 1, very rare : No. 4, common ; No. 6, frequent.

72. HORMOSINA OVICULA, Brady.

Hormosina ovicula, Brady, 1884, Chall. Rep. vol. ix. p. 327, pl. xxxix. figs. 7–9.

Found in Sample No. 1, very rare.

73. HORMOSINA GLOBULIFERA, Brady.

Hormosina globulifera, Brady, 1884, Chall. Rep. vol. ix. p. 326, pl. xxxix. figs. 1–6.

Found in Sample No. 6, very rare.

AMMODISCUS, Reuss.

74. AMMODISCUS INCERTUS (d'Orb.).

Ammodiscus incertus, Brady, 1884, Chall. Rep. vol. ix. p. 330, pl. xxxviii. figs. 1–3.

Found in Sample No. 1, rare ; No. 4, rare ; No. 6, very rare.

75. AMMODISCUS TENUIS, Brady.

Ammodiscus tenuis, Brady, 1884, Chall. Rep. vol. ix. p. 332, pl. xxxviii. figs. 4–6.

Found in Sample No. 6, very rare.

76. AMMODISCUS CHAROIDES (J. & P.).

Ammodiscus charoides, Brady, 1884, Chall. Rep. vol. ix. p. 334, pl. xxxviii. figs. 10–16.

This species is recorded by Parker and Jones from the Red Sea amongst other localities.

Found in Sample No. 1, very rare.

TROCHAMMINA, Parker & Jones.

77. TROCHAMMINA TRULISSATA, Brady.

Trochammina trulissata, Brady, 1884, Chall. Rep. vol. ix. p. 342, pl. xl. figs. 13–16.

Found in Sample No. 1, frequent; No. 4, frequent; No. 6, rare.

WEBBINA, d'Orbigny.

78. WEBBINA CLAVATA, J. & P.

Webbina clavata, Brady, 1884, Chall. Rep. vol. ix. p. 349, pl. xli. figs. 12–16.

Found in Sample No. 4, frequent ; No. 6, very rare.

CYCLAMMINA, Brady.

79. CYCLAMMINA PUSILLA, Brady.

Cyclammina pusilla, Brady, 1884, Chall. Rep. vol. ix. p. 353, pl. xxxvii. figs. 20–23.

Found in Sample No. 1, rare ; No. 3, very rare ; No. 4, frequent ; No. 6, rare.

80. CYCLAMMINA CANCELLATA, Brady.

Cyclammina cancellata, Brady, 1884, Chall. Rep. vol. ix. p. 351, pl. xxxvii. figs. 8–16.

Found in Sample No. 6, rare.

TEXTULARIA, Defrance.

81, TEXTULARIA SAGITTULA, Defrance.

Textularia sagittula, Brady, 1884, Chall. Rep. vol. ix. p. 361, pl. xlii. figs. 17, 18.

Found in Sample No. 4, rare ; No. 5, very rare ; No. 6, very rare.

82. TEXTULARIA SAGITTULA, var. FISTULOSA, Brady.

Textularia sagittula, Brady, 1884, Chall. Rep. vol. ix. p. 362, pl. xlii. figs. 19–22.

Found in Sample No. 1, rare.

83. TEXTULARIA GRAMEN, d'Orb.

Textularia gramen, Brady, 1884, Chall. Rep. vol. ix. p. 365, pl. xliii. figs. 9, 10.

Found in Sample No. 1, frequent.

84. TEXTULARIA LYTHOSTROTUM (Schwager).

Textularia lythostrotum, Schwager, 1866, Novara-Exped., geol. Theil, vol. ii. p. 194, pl. iv. fig. 4 *a–c*.

This species is one of those which have not been recorded from the 'Challenger' gatherings. *T. lythostrotum* was first described from the Pliocene deposits of Kar Nicobar ; it is a very striking form, and is somewhat like *T. gramen* in contour, though with more parallel sides, and the test altogether is very much flattened, the margins being thin and sharp. The surface of the test is usually very rough.

Found in Sample No. 1, frequent ; No. 2, very rare ; No. 4, rare ; No. 6, common.

85. TEXTULARIA CONICA, d'Orb.

Textularia conica, Brady, 1884, Chall. Rep. vol. ix. p. 365, pl. xliii. figs. 13, 14, pl. cxiii. fig. 1 *a, b*.

Found in Sample No. 6, very rare.

86. TEXTULARIA AGGLUTINANS, d'Orb.

Textularia agglutinans, Brady, 1884, Chall. Rep. vol. ix. p. 363, pl. xliii. figs. 1–3.

Found in Sample No. 1, rare ; No. 6, common.

VERNEUILINA, d'Orbigny.

87. VERNEUILINA PYGMÆA (Egger).

Verneuilina pygmæa, Brady, 1884, Chall. Rep. vol. ix. p. 385, pl. xlvii. figs. 4–7.

Found in Sample No. 1, frequent ; No. 4, rare.

88. VERNEUILINA PROPINQUA, Brady.

Verneuilina propinqua, Brady, 1884, Chall. Rep. vol. ix. p. 387, pl. xlvii. figs. 8–14.

Found in Sample No. 1, very rare ; No. 4, rare.

2*

CHRYSALIDINA, d'Orbigny.

89. CHRYSALIDINA DIMORPHA, Brady.

Chrysalidina dimorpha, Brady, 1884, Chall. Rep. vol. ix. p. 388, pl. xlvi. figs. 20, 21.

This species is usually met with in shallow-water deposits, near coral-islands, and also in shore-sands (Brady).

Found in Sample No. 2, rare ; No. 5, very rare.

GAUDRYINA, d'Orbigny.

90. GAUDRYINA PUPOIDES, d'Orb.

Gaudryina pupoides, Brady, 1884, Chall. Rep. vol. ix. p. 378, pl. xlvi. figs. 1-4.

Found in Sample No. 1, rare ; No. 4, very rare ; No. 6, very rare.

91. GAUDRYINA RUGOSA, d'Orb.

Gaudryina rugosa, Brady, 1884, Chall. Rep. vol. ix. p. 381, pl. xlvi. figs. 14-16.

The specimens from the Arabian Sea gatherings are extremely large (about ⅛ inch in length) and well developed.

Found in Sample No. 1, frequent ; No. 3, rare ; No. 4, common ; No. 6, frequent.

92. GAUDRYINA SUBROTUNDATA, Schwager.

Gaudryina subrotundata, Schwager, 1866, Novara-Exped., geol. Theil, vol. ii. p. 198, pl. iv. fig. 9 *a–c*; Brady, 1884, Chall. Rep. vol. ix. p. 380, pl. xlvi. fig. 13 *a–c*.

This form was first described by Schwager from the fossil specimens of the Pliocene beds of Kar Nicobar. It has also been found fossil in Miocene beds of Baden (*G. praelonga*, Karrer). As a recent form it has been recorded from soundings off Culebra Island at 390 fathoms, and off Raine Island at 155 fathoms (Brady).

Found in Sample No. 1, common ; No. 4, frequent ; No. 6, common.

93. GAUDRYINA BACCATA, Schwager.

Gaudryina baccata, Schwager, 1866, Novara-Exped., geol. Theil, vol. ii. p. 200, pl. iv. fig. 12 *a, b*; Brady, 1884, Chall. Rep. vol. ix. p. 379, pl. xlvi. figs. 8-11.

This species is one of those originally described by Dr. Schwager from the Pliocene of Kar Nicobar. It has also been recorded by Dr. Brady from various stations in the Atlantic and Pacific Oceans.

Found in Sample No. 2, very rare.

94. GAUDRYINA SIPHONELLA, Reuss.

Gaudryina siphonella, Brady, 1884, Chall. Rep. vol. ix. p. 382, pl. xlvi. figs. 17-19.

Found in Sample No. 5, very rare ; No. 6, very rare.

VALVULINA, d'Orbigny.

95. VALVULINA CONICA, Parker & Jones.

Valvulina conica, Brady, 1884, Chall. Rep. vol. ix. p. 392, pl. xlix. figs. 15, 16.

Found in Sample No. 1, rare ; No. 6, very rare.

CLAVULINA, d'Orbigny.

96. CLAVULINA COMMUNIS, d'Orbigny.

Clavulina communis, Brady, 1884, Chall. Rep. vol. ix. p. 394, pl. xlviii. figs. 1-13.

Many of the specimens of *C. communis* from the Arabian Sea are greatly elongated, and frequently attain a length of $\frac{1}{8}$ inch.

Found in Sample No. 1, frequent ; No. 3, frequent ; No. 4, common ; No. 5, rare ; No. 6, rare.

97. CLAVULINA PARISIENSIS, d'Orbigny.

Clavulina parisiensis, Brady, 1884, Chall. Rep. vol. ix. p. 395, pl. xlviii. figs. 14-18.

Found in Sample No. 1, very rare.

98. CLAVULINA ANGULARIS, d'Orbigny.

Clavulina angularis, Brady, 1884, Chall. Rep. vol. ix. p. 396 pl. xlviii. figs. 22-24.

Found in Sample No. 1, rare.

BULIMINA, d'Orbigny.

99. BULIMINA OVATA, d'Orbigny.

Bulimina ovata, Brady, 1884, Chall. Rep. vol. ix. p. 400, pl. l. fig. 13 *a, b.*

Found in Sample No. 1, frequent; No. 4, rare; No. 5, very rare.

100. BULIMINA PYRULA, d'Orbigny.

Bulimina pyrula, Brady, 1884, Chall. Rep. vol. ix. p. 399, pl. l. figs. 7-10.

Found in Sample No. 1, frequent; No. 3, rare; No. 5, very rare.

101. BULIMINA ELONGATA, d'Orbigny.

Bulimina elongata, Brady, 1884, Chall. Rep. vol. ix. p. 401, pl. li. figs. 1, 2 ?

Found in Sample No. 1, very rare.

102. BULIMINA PUPOIDES, d'Orbigny.

Bulimina pupoides, Brady, 1884, Chall. Rep. vol. ix. p. 400, pl. l. fig. 15 *a, b.*

Found in Sample No. 1, very rare ; No. 3, very rare.

103. BULIMINA AFFINIS, d'Orbigny.

Bulimina affinis, Brady, 1884, Chall. Rep. vol. ix. p. 400, pl. l. fig. 14 *a, b.*

Found in Sample No. 1, rare.

104. BULIMINA ELEGANS, d'Orbigny.

Bulimina elegans, Brady, 1884, Chall. Rep. vol. ix. p. 398, pl. l. figs. 1–4.

Found in Sample No. 1, very rare.

105. BULIMINA SUBCYLINDRICA, Brady.

Bulimina subcylindrica, Brady, 1884, Chall. Rep. vol. ix. p. 404, pl. l. fig. 16 *a, b.*

Found in Sample No. 1, rare.

106. BULIMINA DECLIVIS, Reuss.

Bulimina declivis, Brady, 1884, Chall. Rep. vol. ix. p. 404, pl. l. fig. 19 *a, b.*

Found in Sample No. 1, very rare.

107. BULIMINA CONTRARIA (Reuss).

Bulimina contraria, Brady, 1884, Chall. Rep. vol. ix. p. 409, pl. liv. fig. 18 *a–c.*

Found in Sample No. 1, frequent; No. 4, rare; No. 5, very rare ; No. 6, frequent.

108. BULIMINA ACULEATA, d'Orbigny.

Bulimina aculeata, Brady, 1884, Chull. Rep. vol. ix. p. 406, pl. li. figs. 7–9.

Found in Sample No. 1, very common ; No. 3, rare ; No. 5, frequent; No. 6, very rare.

109. BULIMINA BUCHIANA, d'Orbigny.

Bulimina buchiana, Brady, 1884, Chall. Rep. vol. ix. p. 407, pl. li. figs. 18, 19.

Found in Sample No. 1, very rare ; No. 4, very rare ; No. 5, very rare.

110. BULIMINA INFLATA, Seguenza.

Bulimina inflata, Brady, 1884, Chall. Rep. vol. ix. p. 406, pl. li. figs. 10–13.

Besides occurring in other Tertiary deposits, Schwager records this species from the Pliocene beds of Kar Nicobar.

Found in Sample No. 1, very rare ; No. 6, very rare.

111. BULIMINA SUBORNATA, Brady.

Bulimina subornata, Brady, 1884, Chall. Rep. vol. ix. p. 402, pl. li. fig. 6 *a, b*.
This rare species was found by Dr. Brady on the *Hyalonema*-ground S. of Japan, at 345 fathoms, and off Aru Island, 800 fathoms.
Found in Sample No. 2, very rare : No. 5, very rare.

112. BULIMINA ROSTRATA, Brady.

Bulimina rostrata, Brady, 1884, Chall. Rep. vol. ix. p. 408, pl. li. figs. 14, 15.
Found in Sample No. 5, rare.

VIRGULINA, d'Orbigny.

113. VIRGULINA SCHREIBERSIANA, Czjzek.

Virgulina schreibersiana, Brady, 1884, Chall. Rep. vol. ix. p. 414, pl. lii. figs. 1–3.
Found in Sample No. 1, very rare ; No. 5, very rare.

114. VIRGULINA SUBSQUAMOSA, Egger.

Virgulina subsquamosa, Brady, 1884, Chall. Rep. vol. ix. p. 415, pl. lii. figs. 7–11.
Found in Sample No. 1, rare.

115. VIRGULINA SUBDEPRESSA, Brady.

Virgulina subdepressa, Brady, 1884, Chall. Rep. vol. ix. p. 416, pl. lii. figs. 14–17.
Found in Sample No. 5, very rare.

BOLIVINA, d'Orbigny.

116. BOLIVINA PUNCTATA, d'Orb.

Bolivina punctata, Brady, 1884, Chall. Rep. vol. ix. p. 417, pl. lii. figs. 18, 19.
Found in Sample No. 1, very rare.

117. BOLIVINA TEXTILARIOIDES, Reuss.

Bolivina textilarioides, Brady, 1884, Chall. Rep. vol. ix. p. 419, pl. lii. figs. 23–25.
Found in Sample No. 5, very rare.

118. BOLIVINA LIMBATA, Brady.

Bolivina limbata, Brady, 1884, Chall. Rep. vol. ix. p. 419, pl. lii. figs. 26–28.
Found in Sample No. 1, very rare ; No. 5, very rare.

119. BOLIVINA NOBILIS, Hantken.

Bolivina nobilis, Brady, 1884, Chall. Rep. vol. ix. p. 424, pl. liii.
figs. 14, 15.
Found in Sample No. 1, very rare ; No. 5, rare.

120. BOLIVINA BEYRICHI, Reuss.

Bolivina beyrichi, Brady, Chall. Rep. vol. ix. p. 422, pl. liii.
fig. 1.

121. BOLIVINA OBSOLETA, Eley.

Bolivina obsoleta, Eley, 1859, Geol. in the Garden, p. 195, pl. ii.
fig. 11, p. 202, pl. viii. fig. 11 c.
Textilaria quadrilatera, Schwager, 1866, Novara-Exped., geol.
Theil, vol. ii. p. 253, pl. vii. fig. 103.
Textularia quadrilatera, Brady, 1884, Chall. Rep. vol. ix. p. 358,
pl. xlii. figs. 8–12.

Dr. Brady records this form (*T. quadrilatera*) from various
stations in the Atlantic and Pacific Oceans, at depths between 350
and 1350 fathoms. That author also suggests that the form
belongs rather to the genus *Bolivina* than to *Textularia*, on
account of the compression of the test, together with the shape of
the aperture. The species *T. quadrilatera* was originally described
as a fossil from the Pliocene of Kar Nicobar ; but was previously
known from the Upper Chalk under the name of *Bolivina obsoleta*.
The characters of both the recent and fossil forms are so nearly
parallel as to satisfy the most critical student.
Found in Sample No. 1, rare ; No. 5, very rare.

122. BOLIVINA ROBUSTA, Brady.

Bolivina robusta, Brady, 1884, Chall. Rep. vol. ix. p. 421, pl. liii.
figs. 7–9.
Found in Sample No. 5, very rare.

123. BOLIVINA ARENOSA, sp. nov. (Plate I. fig. 3 *a, b.*)

Test rhomboidal, compressed ; consisting of about 13 chambers.
Aboral end of test sharply angular. The earlier chambers are
linear, but rapidly increase in breadth. Peripheral edge of the
test somewhat sharp ; and the outline on the lateral aspect
sinuous. Aperture an elongate slit. Test of a pale ochreous
brown colour, finely arenaceous, but with a few coarser particles
interspersed. Length $\frac{1}{23}$ inch (1·08 mm.) ; breadth $\frac{1}{27}$ inch
(·926 mm.).

The above species is an exceptional one in the genus *Bolivina*,
species of that group usually possessing hyaline tests. The
vertical position and slit-like form of the aperture, however,
separate it from the genus *Textularia*.
Found in Sample No. 1, very rare.

PLEUROSTOMELLA, Reuss.

124. PLEUROSTOMELLA SUBNODOSA, Reuss.

Pleurostomella subnodosa, Brady, 1884, Chall. Rep. vol. ix. p. 412, pl. lii. figs. 12, 13.

Found in Sample No. 1, rare.

125. PLEUROSTOMELLA ALTERNANS, Schwager.

Pleurostomella alternans, Schwager, 1866, Novara-Exped., geol. Theil, vol. ii. p. 238, pl. vi. figs. 79, 80 ; Brady, 1884, Chall. Rep. vol. ix. p. 412, pl. li. figs. 22, 23.

This species was originally described by Dr. Schwager from the Pliocene beds of Kar Nicobar. It has been recorded by Dr. Brady from the Ki Islands, S.W. of Papua, 129 fathoms; and S.W. of the Low Archipelago, 2075 fathoms.

Found in Sample No. 6, rare.

Subfamily CASSIDULININÆ.

CASSIDULINA, d'Orbigny.

126. CASSIDULINA MURRHYNA (Schwager).

Sphæroidina murrhyna, Schwager, 1866, Novara-Exped., geol. Theil, vol. ii. p. 250, pl. vii. fig. 97.

This species has not been hitherto recorded from deep-sea soundings. It was described by Schwager from the Pliocene beds of Kar Nicobar.

Found in Sample No. 3, very rare ; No. 4, frequent; No. 5, very rare ; No. 6, frequent.

127. CASSIDULINA CALABRA (Seguenza).

Cassidulina calabra, Brady, 1884, Chall. Rep. vol. ix. p. 431, pl. cxiii. fig. 8 *a-c.*

Recorded from Raine Island, 155 fathoms ; Kandavu, Fiji Ids., 610 fathoms (Brady).

Found in Sample No. 5, rare ; No. 6, common.

128. CASSIDULINA SUBGLOBOSA, Brady.

Cassidulina subglobosa, Brady, 1884, Chall. Rep. vol. ix. p. 430, pl. liv. fig. 17 *a-c.*

Found in Sample No. 1, very rare.

129. CASSIDULINA BRADYI, Norman.

Cassidulina bradyi, Brady, 1884, Chall. Rep. vol. ix. p. 431, pl. liv. figs. 6-10.

Found in Sample No. 1, rare.

130. CASSIDULINA PARKERIANA, Brady.

Cassidulina parkeriana, Brady, 1884, Chall. Rep. vol. ix. p. 432, pl. liv. figs. 11–16.

Hitherto this species has been recorded solely from soundings taken around the islands off the west coast of Patagonia, at depths of 145–175 fathoms (Brady).

Found in Sample No. 6, very rare.

131. CASSIDULINA LÆVIGATA, d'Orbigny.

Cassidulina lævigata, Brady, 1884, Chall. Rep. vol. ix. p. 428, pl. liv. figs. 1–3.

Found in Sample No. 1, rare; No. 5, very rare.

EHRENBERGINA, Reuss.

132. EHRENBERGINA SERRATA, Reuss.

Ehrenbergina serrata, Brady, 1884, Chall. Rep. vol. ix. p. 434, pl. lv. figs. 2–7.

This species was pointed out by Dr. Brady to be not uncommon in recent soundings; he records it off the Azores, 450 fathoms; off the Canaries, 620 fathoms; from the S. Atlantic, 1025 to 2350 fathoms; from the N. Pacific, at 2340 fathoms; and from the S. Pacific, from 150 to 2075 fathoms.

Reuss and Karrer record it from the Miocene beds in the neighbourhood of Vienna.

Found in Sample No. 1, rare; No. 2, frequent; No. 5, frequent.

CHILOSTOMELLA, Reuss.

133. CHILOSTOMELLA OVOIDEA, Reuss.

Chilostomella ovoidea, Brady, 1884, Chall. Rep. vol. ix. p. 436, pl. lv. figs. 12–23.

Found in Sample No. 1, rare; No. 3, very rare.

ALLOMORPHINA, Reuss.

134. ALLOMORPHINA TRIGONA, Reuss.

Allomorphina trigona, Brady, 1884, Chall. Rep. vol. ix. p. 438, pl. lv. figs. 24–26.

This rare foraminifer has previously been recorded from the *Hyalonema*-ground, south of Japan, at 345 fathoms; and off Tahiti, Society Islands, at 620 fathoms (Brady).

Found in Sample No. 6, very rare.

LAGENA, Walker and Boys.

135. LAGENA LÆVIS (Montagu).

Lagena lævis, Brady, 1884, Chall. Rep. vol. ix. p. 455, pl. lvi. figs. 7–14, 30.

Found in Sample No. 1, rare.

136. LAGENA GLOBOSA (Montagu).

Lagena globosa, Brady, 1884, Chall. Rep. vol. ix. p. 452, pl. lvi. figs. 1-3.
Found in Sample No. 6, very rare.

137. LAGENA APICULATA, Reuss.

Lagena apiculata, Brady, 1884, Chall. Rep. vol. ix. p. 453, pl. lvi. figs. 4, 15-18.
Found in Sample No. 1, rare.

138. LAGENA DISTOMA, Parker & Jones.

Lagena distoma, Brady, 1884, Chall. Rep. vol. ix. p. 461, pl. lviii. figs. 11-15.
Found in Sample No. 1, very rare.

139. LAGENA HISPIDA, Reuss.

Lagena hispida, Brady, 1884, Chall. Rep. vol. ix. p. 459, pl. lvii. figs. 1-4, pl. lix. figs. 2, 5.
Found in Sample No. 1, very rare.

140. LAGENA ASPERA, Reuss, var. SPINIFERA, nov.   (Plate I. fig. 4.)

This variety of Reuss's species has the aboral end beset with moderately long spines. Length of the body of the test $\frac{1}{22}$ inch (1·136 mm.).
Found in Sample No. 4, very rare.

141. LAGENA SULCATA (Walker & Jacob).

Lagena sulcata, Brady, 1884, Chall. Rep. vol. ix. p. 462, pl. lvii. figs. 23, 26, 33, 34.
Found in Sample No. 6, very rare.

142. LAGENA GRACILIS, Williamson.

Lagena gracilis, Schwager, 1866, Novara-Exped., geol. Theil, vol. ii. p. 206, pl. iv. fig. 21 *a, b*; Brady, 1884, Chall. Rep. vol. ix. p. 464, pl. lviii. figs. 2, 3, 7-10, 19, 22-24.
This species was also recorded as a fossil from Kar Nicobar.
Found in Sample No. 1, very rare; No. 5, very rare.

143. LAGENA FEILDENIANA, Brady.

Lagena feildeniana, Brady, 1884, Chall. Rep. vol. ix. p. 469, pl. lviii. figs. 38, 39.
Found in Sample No. 3, very rare.

144. LAGENA DESMOPHORA, O. Rymer Jones.

Lagena desmophora, Brady, 1884, Chall. Rep. vol. ix. p. 468, pl. lviii. figs. 42, 43.
Found in Sample No. 1, rare; No. 3, very rare; No. 5, very rare.

145. LAGENA HEXAGONA, Williamson.

Lagena hexagona, Brady, 1884, Chall. Rep. vol. ix. p. 472, pl. lviii. figs. 32, 33.

Found in Sample No. 1, rare.

146. LAGENA MARGINATA (Walker & Jacob).

Lagena marginata, Brady, 1884, Chall. Rep. vol. ix. p. 476, pl. lix. figs. 21–23.

Found in Sample No. 4, very rare ; No. 5, rare ; No. 6, very rare.

147. LAGENA MARGINATA (Walker & Jacob), var. CATENULOSA, nov. (Plate I. fig. 5 *a*, *b*.)

This beautiful variety belongs to the wide-flanged type of *L. marginata*. On the lateral aspect the test is decorated with two or more chain-like borders encircling the bulbous portion. A remarkable feature about this variety is its apiculate base, encompassed, however, within the thin outer flange. The oral extremity of the test is distinctly phialine or lipped, partially closed over with redundant shell-growth, and showing a secondary tubular (true) orifice within. The outer flange is mined by a microscopic boring plant. Length of test $\frac{1}{17}$ inch (1·47 mm.).

One example found in Sample No. 6, very rare.

148. LAGENA SEMINIFORMIS, Schwager.

Lagena seminiformis, Schwager, 1866, Novara-Exped., geol. Theil, vol. ii. p. 208, pl. v. fig. 21 ; Brady, 1884, Chall. Rep. vol. ix. p. 478, pl. lix. figs. 28–30.

This species was recorded as a Pliocene fossil from Kar Nicobar (*Schwager*).

Found in Sample No. 5, very rare.

149. LAGENA LAGENOIDES (Williamson).

Lagena lagenoides, Brady, 1884, Chall. Rep. vol. ix. p. 479, pl. lx. figs. 6, 7, 9, 12, 14.

Found in Sample No. 5, very rare.

150. LAGENA CAPILLOSA (Schwager).

Fissurina capillosa, Schwager, 1866, Novara-Exped., geol. Theil, vol. ii. p. 210, pl. v. fig. 25.

This species is one of those which have not been met with before in deep-sea soundings, and was described from the Pliocene deposits of Kar Nicobar.

Found in Sample No. 6, very rare.

151. LAGENA FIMBRIATA, Brady.

Lagena fimbriata, Brady, 1884, Chall. Rep. vol. ix. p. 486, pl. lx. figs. 26–28.

Found in Sample No. 1, very rare.

152. LAGENA CASTRENSIS, Schwager.

Lagena castrensis, Schwager, 1866, Novara-Exped., geol. Theil, vol. ii. p. 208, pl. v. fig. 22; Brady, 1884, Chall. Rep. vol. ix. p. 485, pl. lx. figs. 1, 2, 3?
This species occurred in the Pliocene deposits of Kar Nicobar.
Found in Sample No. 1, very rare; No. 5, rare.

153. LAGENA STAPHYLLEARIA (Schwager).

Fissurina staphyllearia, Schwager, 1866, Novara-Exped., geol. Theil, vol. ii. p. 209, pl. v. fig. 24.
Lagena staphyllearia, Brady, 1884, Chall. Rep. vol. ix. p. 474, pl. lix. figs. 8–11.
This species occurred in the Pliocene beds of Kar Nicobar.
Found in Sample No. 1, very rare.

154. LAGENA ALVEOLATA, var. SUBSTRIATA, Brady.

Lagena alveolata, var. *substriata,* Brady, 1884, Chall. Rep. vol. ix. p. 488, pl. lx. fig. 34.
This variety was recorded from the Southern Ocean, from 1375 fathoms (Brady).
Found in Sample No. 1, very rare; No. 6, very rare.

155. LAGENA QUADRICOSTULATA, Reuss.

Lagena quadricostulata, Brady, 1884, Chall. Rep. vol. ix. p. 486, pl. lix. figs. 15 and 7?
Found in Sample No. 1, very rare; No. 6, very rare.

156. LAGENA LÆVIGATA (Reuss).

Lagena lævigata, Brady, 1884, Chall. Rep. vol. ix. p. 473, pl. cxiv. fig. 8 *a, b.*
Found in Sample No. 1, very rare.

157. LAGENA ORBIGNYANA (Seguenza).

Lagena orbignyana, Brady, 1884, Chall. Rep. vol. ix. p. 484, pl. lix. figs. 1, 18, 20, 24–26.
Found in Sample No. 1, rare; No. 5, rare; No. 6, frequent.

158. LAGENA FORMOSA, Schwager.

Lagena formosa, Schwager, 1866, Novara-Exped., geol. Theil, vol. ii. p. 206, pl. iv. fig. 19 *a–d*; Brady, 1884, Chall. Rep. vol. ix. p. 480, pl. lx. figs. 10, 18–20, 8?, 17?
This species was originally described from the Pliocene of Kar Nicobar.
Found in Sample No. 1, rare; No. 6, very rare.

159. LAGENA TRIGONO-ORNATA, Brady.

Lagena trigono-ornata, Brady, 1884, Chall. Rep. vol. ix. p. 483, pl. lxi. fig. 14.
Found in Sample No. 1, rare.

160. LAGENA QUADRALATA, Brady.

Lagena quadralata, Brady, 1884, Chall. Rep. vol. ix. p. 464, pl. lxi. fig. 3 *a, b*.

This species has before been recorded from two localities south of Australia at 2600 fathoms, and in the South Atlantic, mid-ocean, 2200 fathoms.
Found in Sample No. 1, very rare.

NODOSARIA, Lamarck.

161. NODOSARIA (DENTALINA) CALOMORPHA, Reuss.

Nodosaria (Dentalina) calomorpha, Brady, 1884, Chall. Rep. vol. ix. p. 497, pl. lxi. figs. 23–27.
Found in Sample No. 5, rare.

162. NODOSARIA RADICULA (Linn.).

Nodosaria radicula, Brady, 1884, Chall. Rep. vol. ix. p. 495, pl. lxi. figs. 28–31.
Found in Sample No. 1, rare ; No. 6, rare.

163. NODOSARIA PYRULA, d'Orbigny.

Nodosaria pyrula, Schwager, 1866, Novara-Exped., geol. Theil, vol. ii. p. 217, pl. v. fig. 38; Brady, 1884, Chall. Rep. vol. ix. p. 497, pl. lxii. figs. 10–12.
This species was also found fossil in the Pliocene deposit of Kar Nicobar.
Found in Sample No. 1, very rare ; No. 5, very rare.

164. NODOSARIA (DENTALINA) FARCIMEN, Reuss (after Soldani).

Nodosaria (Dentalina) farcimen, Brady, 1884, Chall. Rep. vol. ix. p. 498, pl. lxii. figs. 17, 18, woodcut fig. 13 *a–c*.
Found in Sample No. 1, very rare.

165. NODOSARIA (DENTALINA) FILIFORMIS, d'Orbigny.

Nodosaria (Dentalina) filiformis, Brady, 1884, Chall. Rep. vol. ix. p. 500, pl. lxii. figs. 3–5.
Found in Sample No. 1, very rare.

166. NODOSARIA (DENTALINA) ROEMERI (Neugeboren).

Nodosaria (Dentalina) roemeri, Brady, 1884, Chall. Rep. vol. ix. p. 505, pl. lxiii. fig. 1.
Found in Sample No. 1, rare : No. 4, very rare.

167. NODOSARIA (DENTALINA) COMMUNIS, d'Orbigny.

Nodosaria (Dentalina) communis, Brady, 1884, Chall. Rep. vol. ix. p. 504, pl. lxii. figs. 19–22.
Found in Sample No. 1, common ; No. 3, very rare ; No. 5, rare ; No. 6, rare.

168. NODOSARIA (DENTALINA) CONSOBRINA (d'Orbigny).

Nodosaria (Dentalina) consobrina, Brady, 1884, Chall. Rep. vol. ix. p. 501, pl. lxii. figs. 23, 24.

Found in Sample No. 1, rare; No. 2, very rare.

169. NODOSARIA (DENTALINA) INFLEXA, Reuss.

Nodosaria (Dentalina) inflexa, Brady, 1884, Chall. Rep. vol. ix. p. 498, pl. lxii. fig. 9.

Found in Sample No. 5, rare.

170. NODOSARIA OVULATA, Sherborn & Chapman.

Nodosaria ovulata, Sherborn and Chapman, 1886, Journ. Roy. Micr. Soc. ser. 2, vol. vi. p. 747, pl. xiv. fig. 27.

This species was described for the first time from the London Clay of Piccadilly. The specimens from the Arabian Sea agree very closely with the fossil ones.

Found in Sample No. 5, rare.

171. NODOSARIA (DENTALINA) SOLUTA, Reuss.

Nodosaria (Dentalina) soluta, Brady, 1884, Chall. Rep. vol. ix. p. 503, pl. lxii. figs. 13-16.

Found in Sample No. 4, very rare; No. 6, rare.

172. NODOSARIA (DENTALINA) SOLUTA, var. SUBACULEATA, nov. (Plate I. fig. 6.)

"Faintly striate specimen."—*See* Brady, 1884, Chall. Rep. vol. ix. p. 503, pl. lxiv. fig. 28.

This variety differs from the type form in having the basal half of each chamber ornamented with numerous fine prickles, which fade off into faint striæ towards the middle of the bulb. The general contour of the test of this variety agrees with that of the type; and the examples found are well-developed in point of size. Length about ¼ inch (5 mm.).

Dr. Brady has figured a specimen which is undoubtedly referable to the above variety, though it is not so strongly ornamented as are the specimens from the Arabian Sea. For this reason I venture to separate them from the smooth typical forms by a varietal name.

Found in Sample No. 1, rare; No. 3, very rare; No. 4, rare; No. 5, very rare.

173. NODOSARIA (DENTALINA) ACICULA (Lamarck).

Orthocera acicula, Lamarck, 1822, Hist. Anim. sans Vert. vol. vii. p. 594, no. 5.

Dentalina acicula, Sherborn and Chapman, 1886, Journ. Roy. Micr. Soc. ser. 2, vol. vi. p. 751, woodcut fig. 154.

This species is well known as a Tertiary fossil, Lamarck having found it in the Middle Eocene of the Paris Basin; and it is also

characteristic of the London Clay. It does not appear to have
been noticed before as a recent form.
Found in Sample No. 6, very rare.

174. NODOSARIA SCALARIS (Batsch).

Nodosaria scalaris, Brady, 1884, Chall. Rep. vol. ix. p. 510,
pl. lxiii. figs. 28–31.
Found in Sample No. 1, very rare.

175. NODOSARIA SCALARIS (Batsch), var. SEPARANS, Brady.

Nodosaria scalaris, var. *separans*, Brady, 1884, Chall. Rep.
vol. ix. p. 510, pl. lxiv. figs. 16–19.
Found in Sample No. 1, very rare.

176. NODOSARIA (DENTALINA) OBLIQUA (Linn.).

Nodosaria (Dentalina) obliqua, Brady, 1884, Chall. Rep. vol. ix.
p. 513, pl. lxiv. figs. 20–22.
Found in Sample No. 1, very rare.

177. NODOSARIA RAPHANUS (Linn.).

Nodosaria raphanus, Brady, 1884, Chall. Rep. vol. ix. p. 512,
pl. lxiv. figs. 6–10.
Found in Sample No. 1, very rare ; No. 5, rare.

178. NODOSARIA (DENTALINA) ADOLPHINA (d'Orb.).

Dentalina adolphina, d'Orbigny, 1846, Foram. Foss. Vien. p. 51,
pl. ii. figs. 18–20.
Nodosaria adolphina, Schwager, 1866, Novara-Exped., geol.
Theil, vol. ii. p. 235, pl. vi. fig. 72.
This is a well-known Tertiary species, having been recorded
from the London Clay, since d'Orbigny met with it in the Tertiary
strata in the neighbourhood of Vienna. It is interesting to note
that this species was also met with by Dr. Schwager in the Pliocene
deposits of Kar Nicobar. *N. adolphina* does not appear to have
been previously recorded from any deep-sea soundings.
Found in Sample No. 1, very rare ; No. 6, very rare.

179. NODOSARIA (DENTALINA) SUBCANALICULATA (Neugeboren).

Nodosaria (Dentalina) subcanaliculata, Brady, 1884, Chall. Rep.
vol. ix. p. 512, pl. lxiv. figs. 23, 24.
Recorded by Dr. Brady off Tahiti, 420 fathoms. It was also
found fossil in the Miocene of Transylvania.
Found in Sample No. 5, very rare.

180. NODOSARIA (DENTALINA) INTERCELLULARIS ?, Brady.

Nodosaria (Dentalina) intercellularis ?, Brady, 1884, Chall. Rep.
vol. ix. p. 515, pl. lxv. figs. 1–4.
A fragmentary specimen, possibly of this species, was found in
Sample No. 5, very rare.

· RHABDOGONIUM, Reuss.

181. RHABDOGONIUM TRICARINATUM (d'Orb.).

Rhabdogonium tricarinatum, Brady, 1884, Chall. Rep. vol. ix. p. 525, pl. lxvii. figs. 1–3.

Found in Sample No. 5, very rare.

MARGINULINA, d'Orbigny.

182. MARGINULINA GLABRA, d'Orbigny.

Marginulina glabra, Brady, 1884, Chall. Rep. vol. ix. p. 527, pl. lxv. figs. 5, 6.

CRISTELLARIA, Lamarck.

183. CRISTELLARIA ROTULATA (Lamarck).

Cristellaria rotulata, Brady, 1884, Chall. Rep. vol. ix. p. 547, pl. lxix. fig. 13 a, b.

Found in Sample No. 1, very rare ; No. 2, very rare ; No. 6, very rare.

184. CRISTELLARIA CULTRATA (Montfort).

Cristellaria cultrata, Brady, 1884, Chall. Rep. vol. ix. p. 550, pl. lxx. figs. 4–6.

Found in Sample No. 1, frequent; No. 2, rare; No. 6, very rare.

185. CRISTELLARIA ORBICULARIS (d'Orbigny).

Cristellaria orbicularis, Brady, 1884, Chall. Rep. vol. ix, p. 549, pl. lxix. fig. 17.

Found in Sample No. 3, very rare.

186. CRISTELLARIA RENIFORMIS, d'Orbigny.

Cristellaria reniformis, Brady, 1884, Chall. Rep. vol. ix. p. 539, pl. lxx. fig. 3 a, b.

Found in Sample No. 6, very rare.

187. CRISTELLARIA TENUIS (Bornemann).

Cristellaria tenuis, Brady, 1884, Chall. Rep. vol. ix. p. 535, pl. lxvi. figs. 21–23.

Found in Sample No. 1, very rare.

188. CRISTELLARIA OBTUSATA, Reuss, var. SUBALATA, Brady.

Cristellaria obtusata, var. *subalata*, Brady, 1884, Chall. Rep. vol. ix. p. 536, pl. lxvi. figs. 24, 25.

This variety was recorded by Dr. Brady from the N. Atlantic at depths from 130 to 630 fathoms.

Found in Sample No. 1, rare; No. 6, very rare.

189. CRISTELLARIA CREPIDULA (Fichtel & Moll).

Cristellaria crepidula, Brady, 1884, Chall. Rep. vol. ix. p. 542, pl. lxvii. figs. 17, 19, 20, pl. lxviii. figs. 1, 2.

Found in Sample No. 6, very rare.

POLYMORPHINA, d'Orbigny.

190. POLYMORPHINA ANGUSTA, Egger.

Polymorphina angusta, Brady, 1884, Chall. Rep. vol. ix. p. 563, pl. lxxii. figs. 1–3.

Found in Sample No. 1, very rare.

191. POLYMORPHINA OVATA, d'Orbigny.

Polymorphina ovata, Brady, 1884, Chall. Rep. vol. ix. p. 564, pl. lxxii. figs. 7, 8.

This form was found by Dr. Brady off Culebra Island at 390 fathoms.

Found in Sample No. 4, very rare; No. 6, very rare.

192. POLYMORPHINA FUSIFORMIS (Römer).

Globulina fusiformis, Römer, 1838, Neues Jahrb. f. Min. p. 386, pl. iii. fig. 37.

Polymorphina fusiformis, Brady, Parker, and Jones, 1870, Trans. Linn. Soc. Lond. vol. xxvii. p. 219, pl. xxxix. fig. 5 *a–c*, and woodcut *e*.

This species in its typical condition appears to have been hitherto unknown from deep-sea soundings, being previously recorded as a fossil form.

Found in Sample No. 6, very rare.

193. POLYMORPHINA COMMUNIS, d'Orbigny.

Polymorphina communis, Brady, 1884, Chall. Rep. vol. ix. p. 568, pl. lxxii. fig. 19.

Found in Sample No. 1, very rare.

194. POLYMORPHINA SORORIA, Reuss (Fistulose variety).

Polymorphina sororia, Brady, 1884, Chall. Rep. vol. ix. p. 562, pl. lxxiii. fig. 15.

Found in Sample No. 1, very rare.

UVIGERINA, d'Orbigny.

195. UVIGERINA INTERRUPTA, Brady.

Uvigerina interrupta, Brady, 1884, Chall. Rep. vol. ix. p. 580, pl. lxxv. figs. 12–14.

Found in Sample No. 1, very rare.

196. UVIGERINA TENUISTRIATA, Reuss.

Uvigerina tenuistriata, Brady, 1884, Chall. Rep. vol. ix. p. 574, pl. lxxiv. figs. 4–7.
Found in Sample No. 1, very rare; No. 2, rare; No. 5, very rare.

197. UVIGERINA PYGMÆA, d'Orbigny.

Uvigerina pygmæa, Brady, 1884, Chall. Rep. vol. ix. p. 575, pl. lxxiv. figs. 11–14.
Found in Sample No. 1, rare; No. 5, very rare.

198. UVIGERINA ACULEATA, d'Orbigny.

Uvigerina aculeata, Brady, 1884, Chall. Rep, vol. ix. p. 578, pl. lxxv. figs. 1, 2.
Found in Sample No. 1, common; No. 6, frequent.

199. UVIGERINA ANGULOSA, Williamson.

Uvigerina angulosa, Brady, 1884, Chall. Rep. vol. ix. p. 576, pl. lxxiv. figs. 15–18.
Found in Sample No. 2, rare; No. 5, rare.

200. UVIGERINA ANGULOSA, Williamson, var. SPINIPES, Brady.

Uvigerina angulosa, var. *spinipes*, Brady, 1884, Chall. Rep. vol. ix. p. 577, pl. lxxiv. figs. 19, 20.
This variety has been recorded by Dr. Brady from one locality only—Nightingale Island, Tristan d'Acunha, 100–150 fathoms.
Found in Sample No. 2, very rare.

201. UVIGERINA ASPERULA, Czjzek.

Uvigerina asperula, Brady, 1884, Chall. Rep. vol. ix. p. 578, pl. lxxv. figs. 6–8.
This form was figured by Dr. Schwager from the Pliocene deposits of Kar Nicobar under the name of *Uvigerina hispida*.
Found in Sample No. 2, very rare; No. 3, frequent; No. 4, very rare; No. 5, rare; No. 6, frequent.

202. UVIGERINA ASPERULA, Czjzek, var. AMPULLACEA, Brady.

Uvigerina asperula, var. *ampullacea*, Brady, 1884, Chall. Rep. vol. ix. p. 579, pl. lxxv. figs. 10, 11.
Found in Sample No. 5, rare; No. 6, very rare.

203. UVIGERINA SCHWAGERI, Brady.

Uvigerina schwageri, Brady, 1884, Chall. Rep. vol. ix. p. 575, pl. lxxiv. figs. 8–10.
This species was recorded by Dr. Brady from Kandavu, Fiji Islands, 210 fathoms; Raine Island, Torres Strait, 155 fathoms; and off the Philippine Islands, 95 fathoms.
Found in Sample No. 2, frequent.

3*

204. UVIGERINA CANARIENSIS, d'Orbigny.

Uvigerina canariensis, Brady, 1884, Chall. Rep. vol. ix. p. 573, pl. lxxiv. figs. 1–3.

This species was obtained by Dr. Schwager from the Pliocene beds of Kar Nicobar, and figured under the name of *Uvigerina proboscidea*.

Found in Sample No. 2, very rare ; No. 5, rare.

205. UVIGERINA BRUNNENSIS, Karrer.

Uvigerina brunnensis, Brady, 1884, Chall. Rep. vol. ix. p. 577, pl. lxxv. figs. 4, 5.

This form has before occurred off Christmas Harbour, Kerguelen Island, 120 fathoms ; and on the western shores of Patagonia, 245 fathoms (Brady).

Found in Sample No. 4, very rare.

SAGRINA (d'Orbigny), Parker & Jones.

206. SAGRINA COLUMELLARIS, Brady.

Sagrina columellaris, Brady, 1884, Chall. Rep. vol. ix. p. 581, pl. lxxv. figs. 15–17.

Found in Sample No. 1, very rare.

RAMULINA, Rupert Jones.

(There is some probability of this organism belonging to a form of *Polymorphina*.)

207. RAMULINA GLOBULIFERA, Brady.

Ramulina globulifera, Brady, 1884, Chall. Rep. vol. ix. p. 587, pl. lxxvi. figs. 22–28.

Found in Sample No. 5, very rare.

GLOBIGERINA, d'Orbigny.

208. GLOBIGERINA BULLOIDES, d'Orbigny.

Globigerina bulloides, Brady, 1884, Chall. Rep. vol. ix. p. 593, pls. lxxvii., lxxix. figs. 3–7.

Found in Sample No. 1, common ; No. 2, common ; No. 4, frequent ; No. 5, rare ; No. 6, common.

209. GLOBIGERINA BULLOIDES, d'Orb., var. TRILOBA, Reuss.

Globigerina bulloides, var. *triloba*, Brady, 1884, Chall. Rep. vol. ix. p. 595, pl. lxxix. figs. 1, 2, pl. lxxxi. figs. 2, 3.

Found in Sample No. 2, very rare; No. 3, rare; No. 5, very rare ; No. 6, very rare.

210. GLOBIGERINA DUBIA, Egger.

Globigerina dubia, Brady, 1884, Chall. Rep. vol. ix. p. 595, pl. lxxix. fig. 17 *a–c*.

Found in Sample No. 3, rare; No. 5, very rare; No. 6, very rare.

211. GLOBIGERINA RUBRA, d'Orbigny.

Globigerina rubra, Brady, 1884, Chall. Rep. vol. ix. p. 602, pl. lxxix. figs. 11–16.

Found in Sample No. 1, very rare; No. 2, rare; No. 4, rare; No. 6, rare.

212. GLOBIGERINA CRETACEA, d'Orbigny.

Globigerina cretacea, Brady, 1884, Chall. Rep. vol. ix. p. 596, pl. lxxxii. fig. 10 *a–c*.

Found in Sample No. 1, common; No. 2, common; No. 3, very rare; No. 4, frequent; No. 5, rare; No. 6, frequent.

213. GLOBIGERINA CONGLOBATA, Brady.

Globigerina conglobata, Brady, 1884, Chall. Rep. vol. ix. p. 603, pl. lxxx. figs. 1–5; pl. lxxxii. fig. 5.

Found in Sample No. 1, common; No. 2, common; No. 4, common; No. 5, rare; No. 6, frequent.

214. GLOBIGERINA ÆQUILATERALIS, Brady.

Globigerina æquilateralis, Brady, 1884, Chall. Rep. vol. ix. p. 605, pl. lxxx. figs. 18–21.

Found in Sample No. 1, very common; No. 2, rare; No. 3, very rare; No. 4, common; No. 5, rare; No. 6, common.

215. GLOBIGERINA SACCULIFERA, Brady.

Globigerina sacculifera, Brady, 1884, Chall. Rep. vol. ix. p. 604, pl. lxxx. figs. 11–17, pl. lxxxii. fig. 4.

Found in Sample No. 1, very common; No. 2, rare; No. 3, frequent; No. 4, common; No. 5, rare; No. 6, rare.

216. GLOBIGERINA DIGITATA, Brady.

Globigerina digitata, Brady, 1884, Chall. Rep. vol. ix. p. 599, pl. lxxx. figs. 6–10, pl. lxxxii. figs. 6, 7.

Dr. Brady states that this species was found at three stations in the South Atlantic and at six in the South Pacific, and also near the Ki Islands in the Eastern Archipelago at 580 fathoms, the latter place being the only one at which it was found in any abundance.

It is therefore interesting to note the occurrence of *G. digitata* in the Arabian Sea, where this fantastic species is common and in some instances remarkably developed.

Found in Sample No. 1, very common; No. 4, very common; No. 5, frequent; No. 6, very rare.

ORBULINA, d'Orbigny.

217: ORBULINA UNIVERSA, d'Orbigny.

Orbulina universa, Brady, Chall. Rep. vol. ix. p. 608, pl. lxxviii., pl. lxxxi. figs. 8–26, pl. lxxxii. figs. 1–3.

Found in Sample No. 1, very common; No. 3, rare; No. 4, common: No. 5, common; No. 6, very common.

HASTIGERINA, Wyv. Thomson.

218. HASTIGERINA PELAGICA (d'Orbigny).

Hastigerina pelagica, Brady, 1884, Chall. Rep. vol. ix. p. 613, pl. lxxxiii. figs. 1–8.

Found in Sample No. 6, very rare.

PULLENIA, Parker & Jones.

219. PULLENIA OBLIQUILOCULATA, Parker & Jones.

Pullenia obliquiloculata, Brady, 1884, Chall. Rep. vol. ix. p. 618, pl. lxxxiv. figs. 16–20.

Found in Sample No. 1, very common; No. 2, frequent; No. 3, rare; No. 4, common; No. 5, common; No. 6, frequent.

220. PULLENIA SPHÆROIDES (d'Orbigny).

Pullenia sphæroides, Brady, 1884, Chall. Rep. vol. ix. p. 615, pl. lxxxiv. figs. 12, 13.

Found in Sample No. 5, rare; No. 6, rare.

221. PULLENIA QUINQUELOBA, Reuss.

Pullenia quinqueloba, Brady, 1884, Chall. Rep. vol. ix. p. 617, pl. lxxxiv. figs. 14, 15.

Found in Sample No. 6, very rare.

SPHÆROIDINA, d'Orbigny.

222. SPHÆROIDINA BULLOIDES, d'Orbigny.

Sphæroidina bulloides, Brady, 1884, Chall. Rep. vol. ix. p. 620, pl. lxxxiv. figs. 1–7.

Found in Sample No. 1, very rare; No. 3, very rare; No. 4, rare; No. 6, rare.

223. SPHÆROIDINA DEHISCENS, Parker & Jones.

Globigerina seminulina, Schwager, 1866, Novara Exped., geol. Theil, vol. ii. p. 256, pl. vii. fig. 112.

Sphæroidina dehiscens, Brady, 1884, Chall. Rep. vol. ix. p. 621, pl. lxxxiv. figs. 8–11.

This species was recorded by Schwager from the Pliocene of Kar Nicobar under the name of *Globigerina seminulina.*

Found in Sample No. 1, common; No. 3, rare; No. 4, frequent; No. 6, rare.

CANDEINA, d'Orbigny.

224. CANDEINA NITIDA, d'Orbigny.

Candeina nitida, Brady, 1884, Chall. Rep. vol. ix. p. 622, pl. lxxxii. figs. 13-20.

Found in Sample No. 5, very rare.

CYMBALOPORA, Hagenow.

225. CYMBALOPORA POEYI (d'Orbigny).

Cymbalopora poeyi, Brady, 1884, Chall. Rep. vol. ix. p. 636, pl. cii. fig. 13 *a-c*.

Found in Sample No. 1, very rare ; No. 2, rare.

226. CYMBALOPORA (TRETOMPHALUS) BULLOIDES (d'Orb.).

Cymbalopora (Tretomphalus) bulloides, Brady, Chall. Rep. vol. ix. p. 638, pl. cii. figs. 7-12.

Found in Sample No. 1, very rare.

DISCORBINA, Parker & Jones.

227. DISCORBINA VENTRICOSA, Brady.

Discorbina ventricosa, Brady, 1884, Chall. Rep. vol. ix. p. 654, pl. xci. fig. 7 *a-c*.

Found in Sample No. 1, very rare.

228. DISCORBINA PARISIENSIS (d'Orbigny).

Discorbina parisiensis, Brady, 1884, Chall. Rep. vol. ix. p. 648, pl. xc. figs. 5, 6, 9-12.

Found in Sample No. 1, very rare.

229. DISCORBINA ROSACEA (d'Orbigny).

Discorbina rosacea, Brady, 1884, Chall. Rep. vol. ix. p. 644, pl. lxxxvii. figs. 1, 4.

Found in Sample No. 5, very rare ; No. 6, very rare.

230. DISCORBINA RUGOSA (d'Orbigny).

Discorbina rugosa, Brady, 1884, Chall. Rep. vol. ix. p. 652, pl. lxxxvii. fig. 3 *a-c*, pl. xci. fig. 4 *a-c*.

Found in Sample No. 6, rare.

PLANORBULINA, d'Orbigny.

231. PLANORBULINA ACERVALIS, Brady.

Planorbulina acervalis, Brady, 1884, Chall. Rep. vol. ix. p. 657, pl. xcii. fig. 4.

Of this species Dr. Brady remarks, " not uncommon amongst the Islands of the Pacific, in the Indian Ocean, and the Red Sea."

Found in Sample No. 1, very rare ; No. 2, very rare.

232. PLANORBULINA LARVATA, Parker & Jones.

Planorbulina larvata, Brady, 1884, Chall. Rep. vol. ix. p. 658, pl. xcii. figs. 5, 6.

' Found in Sample No. 6, very rare.

TRUNCATULINA, d'Orbigny.

233. TRUNCATULINA LOBATULA (Walker & Jacob).

Truncatulina lobatula, Brady, 1884, Chall. Rep. vol. ix. p. 660, pl. xcii. fig. 10, pl. xciii. figs. 1, 4, 5, pl. xcv. figs. 4, 5.

Found in Sample No. 1, rare ; No. 5, very rare ; No. 6, rare.

234. TRUNCATULINA WUELLERSTORFI (Schwager).

Anomalina wuellerstorfi, Schwager, 1866, Novara-Exped., geol. Theil, vol. ii. p. 258, pl. vii. figs. 105, 107.

Truncatulina wuellerstorfi, Brady, 1884, Chall. Rep. vol. ix. p. 662, pl. xciii. figs. 8, 9.

This species was originally described from fossil specimens from the Pliocene of Kar Nicobar.

Found in Sample No. 1, very common ; No. 3, common ; No. 4, common ; No. 5, rare ; No. 6, rare.

235. TRUNCATULINA PYGMÆA, Hantken.

Truncatulina pygmæa, Brady, 1884, Chall. Rep. vol. ix. p. 666, pl. xcv. figs. 9, 10.

Found in Sample No. 1, rare ; No. 5, rare.

236. TRUNCATULINA UNGERIANA (d'Orbigny).

Truncatulina ungeriana, Brady, 1884, Chall. Rep. vol. ix. p. 664, pl. xciv. fig. 9 *a–c*.

Found in Sample No. 1, frequent ; No. 2, rare ; No. 3, rare ; No. 4, common ; No. 5, rare ; No. 6, frequent.

237. TRUNCATULINA HAIDINGERII (d'Orbigny).

Truncatulina haidingerii, Brady, 1884, Chall. Rep. vol. ix. p. 663, pl. xcv. fig. 7 *a–c*.

Found in Sample No. 1, rare ; No. 5, rare ; No. 6, very rare.

238. TRUNCATULINA ROBERTSONIANA, Brady.

Truncatulina robertsoniana, Brady, 1884, Chall. Rep. vol. ix. p. 664, pl. xcv. fig. 4 *a–c*.

Found in Sample No. 1, rare ; No. 6, very rare.

239. TRUNCATULINA DUTEMPLEI, d'Orbigny.

Truncatulina dutemplei, Brady, 1884, Chall. Rep. vol. ix. p. 665, pl. xcv. fig. 5 *a–c*.

Found in Sample No. 1, rare.

240. TRUNCATULINA PRÆCINCTA (Karrer).

Truncatulina præcincta, Brady, 1884, Chall. Rep. vol. ix. p. 667, pl. xcv. figs. 1–3.

This species affects the neighbourhood of coral-reefs and tropical areas; and occurs, amongst other places, in the Red Sea at 30 fathoms (Brady).
Found in Sample No. 2, rare.

241. TRUNCATULINA AKNERIANA (d'Orbigny).

Truncatulina akneriana, Brady, 1884, Chall. Rep. vol. ix. p. 663, pl. xciv. fig. 8 *a–c*.
Found in Sample No. 3, very rare.

242. TRUNCATULINA TENERA, Brady.

Truncatulina tenera, Brady, 1884, Chall. Rep. vol. ix. p. 665, pl. xcv. fig. 11 *a–c*.
Found in Sample No. 4, rare; No. 5, very rare; No. 6, very rare.

243. TRUNCATULINA CULTER (Parker & Jones).

Truncatulina culter, Brady, 1884, Chall. Rep. vol. ix. p. 668, pl. xcvi. fig. 3 *a–c*.
Anomalina bengalensis, Schwager, 1866, Novara-Exped., geol. Theil, vol. ii. p. 259, pl. vii. fig. 111.
This species was recorded under the latter name by Dr. Schwager from the Pliocene of Kar Nicobar.
Found in Sample No. 6, very rare.

ANOMALINA, d'Orbigny.

244. ANOMALINA GROSSERUGOSA (Gümbel).

Anomalina grosserugosa, Brady, 1884, Chall. Rep. vol. ix. p. 673, pl. xciv. figs. 4, 5.
Found in Sample No. 1, rare; No. 6, frequent.

245. ANOMALINA POLYMORPHA, Costa.

Anomalina polymorpha, Brady, 1884, Chall. Rep. vol. ix. p. 676, pl. xcvii. figs. 3–7.
Found in Sample No. 2, very rare.

PULVINULINA, Parker & Jones.

246. PULVINULINA REPANDA (F. & M.), var. CONCAMERATA, (Montagu).

Pulvinulina repanda, var. *concamerata*, Brady, 1884, Chall. Rep. vol. ix. p. 685, pl. civ. fig. 19 *a–c*.
Found in Sample No. 1, very rare.

247. PULVINULINA ELEGANS (d'Orbigny).

Pulvinulina elegans, Brady, 1884, Chall. Rep. vol. ix. p. 699, pl. cv. figs. 4-6.

One of the specimens found has a carinate edge similar to that in fig. 6 in the 'Challenger' Report.

Found in Sample No. 1, common; No. 3, frequent; No. 4, very common; No. 5, very common; No. 6, common.

248. PULVINULINA PARTSCHIANA (d'Orbigny).

Pulvinulina partschiana, Brady, 1884, Chall. Rep. vol. ix. p. 699, pl. cv. fig. 3 *a-c,* woodcut fig. 21.

This form, which represents the deep-water variety of *P. elegans,* is, as might be supposed, not well-represented in these soundings, as regards well-defined specimens, though transitional forms are frequent.

Found in Sample No. 1, rare : No. 4, rare ; No. 5, rare : No. 6, rare.

249. PULVINULINA MENARDII (d'Orbigny).

Pulvinulina menardii, Brady, 1884, Chall. Rep. vol. ix. p. 690, pl. ciii. figs. 1, 2.

Found in Sample No. 1, very common ; No. 2, common ; No. 4, common; No. 5, frequent; No. 6, common.

250. PULVINULINA MENARDII (d'Orb.), var. FIMBRIATA, Brady.

Pulvinulina menardii, var. *fimbriata,* Brady, 1884, Chall. Rep. vol. ix. p. 691, pl. ciii. fig. 3 *a, b.*

Found in Sample No. 1, very rare ; No. 4, rare.

251. PULVINULINA CANARIENSIS (d'Orbigny).

Pulvinulina canariensis, Brady, 1884, Chall. Rep. vol. ix. p. 692, pl. ciii. figs. 8-10.

Found in Sample No. 1, rare ; No. 5, very rare ; No. 6, rare.

252. PULVINULINA TUMIDA, Brady.

Pulvinulina tumida, Brady, 1884, Chall. Rep. vol. ix. p. 692, pl. ciii. figs. 4-6.

Found in Sample No. 1, frequent ; No. 4, rare ; No. 6, frequent.

253. PULVINULINA MICHELINIANA (d'Orbigny).

Pulvinulina micheliniana, Brady, 1884, Chall. Rep. vol. ix. p. 694, pl. civ. figs. 1, 2.

Found in Sample No. 1, rare; No. 3, common ; No. 6, very rare.

254. PULVINULINA PAUPERATA, Parker & Jones.

Pulvinulina pauperata, Brady, 1884, Chall. Rep. vol. ix. p. 696, pl. civ. figs. 3-11.

The specimens from the Arabian Sea are extremely fine and characteristic.

Found in Sample No. 1, rare; No. 3, rare; No. 4, frequent;
No. 6, rare.

255. PULVINULINA AURICULA (Fichtel & Moll).

Pulvinulina auricula, Brady, 1884, Chall. Rep. vol. ix. p. 688,
pl. cvi. fig. 5 *a–c*.
Found in Sample No. 1, rare; No. 2, very rare.

256. PULVINULINA OBLONGA (Williamson).

Pulvinulina oblonga, Brady, 1884, Chall. Rep. vol. ix. p. 688,
pl. cvi. fig. 4 *a–c*.
Found in Sample No. 1, rare.

257. PULVINULINA PUNCTULATA (d'Orbigny).

Pulvinulina punctulata, Brady, 1884, Chall. Rep. vol. ix. p. 685,
pl. civ. fig. 17 *a–c*.
Found in Sample No. 1, very rare.

258. PULVINULINA EXIGUA, Brady.

Pulvinulina exigua, Brady, 1884, Chall. Rep. vol. ix. p. 696,
pl. ciii. figs. 13, 14.
Found in Sample No. 2, very rare; No. 5, very rare; No. 6,
very rare.

259. PULVINULINA KARSTENI (Reuss).

Pulvinulina karsteni, Brady, 1884, Chall. Rep. vol. ix. p. 698,
pl. cv. figs. 8, 9.

ROTALIA, Lamarck.

260. ROTALIA ORBICULARIS, d'Orbigny.

Rotalia orbicularis, Brady, 1884, Chall. Rep. vol. ix. p. 706,
pl. cvii. fig. 5, pl. cxv. fig. 6.
Found in Sample No. 1, very rare; No. 5, rare; No. 6, very
rare.

261. ROTALIA CALCAR (d'Orbigny).

Rotalia calcar, Brady, 1884, Chall. Rep. vol. ix. p. 709, pl. cviii.
figs. 3, 4?
Found in Sample No. 1, common.

262. ROTALIA SOLDANII, d'Orbigny.

Rotalia soldanii, Brady, 1884, Chall. Rep. vol. ix. p. 706, pl. cvii.
figs. 6, 7.
Found in Sample No. 1, common; No. 3, rare; No. 4, rare;
No. 5, rare; No. 6, frequent.

263. ROTALIA BROECKHIANA, Karrer.

Rotalia broeckhiana, Brady, 1884, Chall. Rep. vol. ix. p. 705,
pl. cvii. fig. 4 *a–c*.

This species has previously been recorded off the Ki Islands, at
a depth of 580 fathoms (Brady).
Found in Sample No. 3, very rare; No. 5, very rare; No. 6,
common.

CALCARINA, d'Orbigny.

264. CALCARINA HISPIDA, Brady.

Calcarina hispida, Brady, 1884, Chall. Rep. vol. ix. p. 713,
pl. cviii. figs. 8, 9.
Found in Sample No. 2, rare.

265. CALCARINA NICOBARENSIS, Schwager.

Calcarina nicobarensis, Schwager, 1866, Novara-Exped., geol.
Theil, vol. ii. p. 261, pl. vii. fig. 114, and fig. 3 (shell-section).
This species was described by Dr. Schwager from the Pliocene
beds of Kar Nicobar, and is here recorded for the first time as a
recent form from the Arabian Sea.
Found in Sample No. 1, very common ; No. 2, rare.

266. CALCARINA DEFRANCEI, d'Orbigny.

Calcarina defrancei, Brady, 1884, Chall. Rep. vol. ix. p. 714,
pl. cviii. fig. 6 *a–c*.
Found in Sample No. 1, very rare.

GYPSINA, Carter.

267. GYPSINA GLOBULUS (Reuss).

Gypsina globulus, Brady, 1884, Chall. Rep. vol. ix. p. 717,
pl. ci. fig. 8.
This species was found in the coral-sands of Sample No. 2,
very rare.

POLYTREMA, Risso.

268. POLYTREMA MINIACEUM (Linné).

Polytrema miniaceum, Brady, 1884, Chall. Rep. vol. ix. p. 721,
pl. c. figs. 5–9, pl. ci. fig. 1.
Found in Sample No. 1, rare.

NONIONINA, d'Orbigny.

269. NONIONINA UMBILICATULA (Montagu).

Nonionina umbilicatula, Brady, 1884, Chall. Rep. vol. ix. p. 726,
pl. cix. figs. 8, 9.
Found in Sample No. 1, rare ; No. 5, frequent ; No. 6, rare.

270. NONIONINA POMPILIOIDES (Fichtel & Moll).

Nonionina pompilioides, Brady, 1884, Chall. Rep. vol. ix. p. 727, pl. cix. figs. 10, 11.

Found in Sample No. 1, very rare; No. 5, very rare; No. 6, rare.

POLYSTOMELLA, Lamarck.

271. POLYSTOMELLA CRISPA (Linné).

Polystomella crispa, Brady, 1884, Chall. Rep. vol. ix. p. 736, pl. cx. figs. 6, 7.

Found in Sample No. 1, rare; No. 2, very rare.

AMPHISTEGINA, d'Orbigny.

272. AMPHISTEGINA LESSONII, d'Orbigny.

Amphistegina lessonii, Brady, 1884, Chall. Rep. vol. ix. p. 740, pl. cxi. figs. 1–7.

Found in Sample No. 1, frequent; No. 2, common; No. 4, rare.

273. AMPHISTEGINA RADIATA (Fichtel & Moll).   (Plate I. figs. 8, 9, 10, 12.)

Nautilus radiatus, Fichtel & Moll, 1803, Test. Micr. p. 58, pl. viii. figs. *a–d*.

The above species was described by Fichtel and Moll from specimens found in sea-sand from the interior of shells from the Red Sea.

Profs. Parker and Jones remark [1] on this form as follows :—

" This is a small, smooth, lenticular *Nummulina*, about 1 line in diameter ; marked with twenty-four radiating, translucent, septal lines, slightly sinuous, with an open sigmoid flexure, which extends from the periphery to the umbonal centre, and as many intermediate, short, parallel septal lines towards the peripherical margin. These indicate altogether nearly fifty chambers in the outer whorl, the lateral lobes of which, in passing towards the umbonal centre, interfere with each other, leaving only indications of half as many elongate, triangular, sinuous, umbilical lobes."

Having this opportunity of examining the very fine specimens referable to the above species, which were found in the above-mentioned (no. 2) coral-deposits of the Laccadives, I prepared slices of the tests, both median and transverse, in the hope of finding additional evidence regarding the affinities of the species. This was considered necessary, especially since the recent examples of *Nummulina* appear to have been hitherto somewhat neglected.

In the first place, the specimens of *Amphistegina radiata* which occur in the Laccadive Island deposits are inequilateral in transverse

---

[1] " The Nomenclature of the Foraminifera, Part III.," Ann. & Mag. Nat. Hist. ser. 3, vol. v. [1860] pp. 105, 106.

section, the umbonal centre being more prominent on one side than the other. This fact points to the tendency of this species to increase in an oblique or turbinoid spiral, such as is shown in all undoubted *Amphisteginæ* and not on the Nummuline plan. I venture to suggest that the peripheral figure of this form, as originally given by Fichtel and Moll[1], is too symmetrically drawn, and it is easy to conceive how such a slight degree of asymmetry would be overlooked without the accompaniment of carefully prepared sections of the test.

Another feature, moreover, brought out in the transverse sections of the test, and which helps to strengthen the evidence in favour of this form belonging to the genus *Amphistegina*, is the existence of the characteristic double cone-shaped non-tubulate portions of the test which form its central axis in transverse section (see Plate 1. fig. 9).

Whilst examining the median sections of *A. radiata*, the presence of true interseptal canals with many branchlets was detected (see fig. 10). In his 'Introduction to the Study of the Foraminifera,' Dr. Carpenter describes the various characters which distinguish forms of the genus *Amphistegina*, and of which a summary and comparison with the Rotaline type is given at p. 246. Here it is remarked that the "singleness of the septal lamellæ is a most important additional link of affinity" to the group of the Rotalines. This statement, which may have been made through the examination of non-typical specimens, caused me some doubt as to the validity of the claim of *A. radiata* to the Amphistegine group. Upon preparing sections of typical specimens of *Amphistegina hauerina* from the Vienna Basin, which I possess through the kindness of Professor T. Rupert Jones, I found the same well-developed canal-system existing in the fossil forms (see fig. 11), of the true position of which as *Amphisteginæ* there can be no question, as were seen in the recent specimens of *A. radiata*.

Therefore that apparently serious objection was satisfactorily removed, and, at the same time, additional facts were obtained, which show that, as far as the shell-structure is concerned, *Amphistegina* is as highly advanced in differential characters as is the shell of *Nummulina*. The only difference therefore that appears to exist between ordinary *Amphisteginæ* of the *A. lessonii* type (including *A. hauerina*) and the recent *A. radiata* is the remarkable modification of the segments in the former type of the outer layer on the inferior side of the test giving rise to the "astral lobes."

The transverse sections of the *Amphisteginæ* generally, if taken accurately through the middle of the shell, exhibit the large spherical primordial chamber with the succeeding more or less ovoid one. I especially mention this fact since several examples of the young tests of *A. radiata* have occurred in the peripheral whorls of adult specimens of that species, and are seen in both median and

[1] *Op. cit.* pl. viii. fig. *d*.

transverse sections of the tests (see fig. 12). In all cases those
observed consisted of two chambers, and they are exactly com-
parable in shape with the early chambers of the adult specimens,
which, by the way, belong to the megalospheric type of growth.
In a paper on "*L'Amphistegina* del calcare lenticolare di Par-
lascio"[1], Dr. G. A. de Amicis described a form of *Amphistegina*
formerly named *Nummulites targionii* by Professor Meneghini, but
which is shown by Dr. de Amicis to be a true *Amphistegina*.
It appears to approach very closely in its general characters to
A. radiata, which it especially resembles in the subdivision of
the peripheral margins of the successive layers seen in transverse
section[2].

The specimens of *A. radiata* found in the Arabian Sea average
⅛ of an inch in diameter, and are thus somewhat larger than the
specimens originally described by Fichtel and Moll. The usual
number of principal septa appearing on the surface of full-grown
individuals is from eighteen to twenty, and these septa often show
a tendency to bifurcate towards the periphery. The surface of
the test is tolerably smooth, and the septal lines are marked out in
clear transparent shell-matter, while the rest of the test is of a
creamy-white colour.

Incidentally I may mention that although the form which has
been referred to as *Nummulina cumingii* by Drs. Carpenter and
Brady has not occurred in these deposits. I have no doubt that,
as Professor Rupert Jones has already suggested to me, that
species is more properly referable to the "*Nautilus venosus*" of
Fichtel and Moll, and should perhaps stand as *Nummulina venosa*
(F. & M.). It is open to some question, however, whether it is
a truly Nummuline form, since some published drawings of the
species show a decidedly inequilateral growth, and in point of fact
a series of specimens may show all gradations into *Operculina*.

Amphistegina radiata was found only in Sample No. 2, in which
it was common.

OPERCULINA, d'Orbigny.

274. OPERCULINA COMPLANATA (Defrance).

Lenticulites complanata, Defrance, 1822, Dict. Sci. Nat. vol. xxv.
p. 453.
Operculina arabica, Carter, 1853, Journ. Roy. Asiatic Soc.,
Bombay Branch, vol. iv. p. 437, pl. xviii.
Operculina complanata, Brady, 1884, Chall. Rep. vol. ix. p. 743,
pl. cxii. figs. 3–5, 8.

The above species was described under the name of *O. arabica*
by H. J. Carter, who obtained his specimens off the south-east
coast of Arabia.

Found in Sample No. 1, frequent.

[1] 1885. Processi verbali della Società Toscana di Scienze Naturali.
[2] See also De Amicis, 1886, "Il Calcare ad *Amphistegina* nella Provincia di
Pisa," Atti Soc. Toscana Sci. Nat., Memorie, vol. vii. pl. xi. figs. 1, 3, 6, 7.

275. OPERCULINA COMPLANATA (Defr.), var. GRANULOSA, Leymerie.

Operculina complanata, var. *granulosa*, Brady, 1884, Chall. Rep. vol. ix. p. 743, pl. cxii. figs. 6, 7, 9, 10.
Found in Sample No. 2, rare.

HETEROSTEGINA, d'Orbigny.

276. HETEROSTEGINA DEPRESSA, d'Orbigny.

Heterostegina depressa, Brady, 1884, Chall. Rep. vol. ix. p. 746, pl. cxii. figs. 14–20.
Found in Sample No. 1, very rare ; No. 2, frequent.

CYCLOCLYPEUS, Carpenter.

277. CYCLOCLYPEUS GUEMBELIANUS, Brady.

Cycloclypeus guembelianus, Brady, 1884, Chall. Rep. vol. ix. p. 751, pl. cxi. fig. 8 *a, b.*
This species has been previously found off Kandavu, Fiji Islands, at 210 fathoms.
Found in Sample No. 1, very rare.

Species and Varieties.	Samples.					
	1.	2.	3.	4.	5.	6.
Family MILIOLIDÆ.						
Subfamily NUBECULARIINÆ.						
1. Nubecularia lucifuga, *Defrance*	v. r.					
Subfamily MILIOLININÆ.						
2. Biloculina depressa, *d'Orbigny*	c.	...	f.	c.	r.	f.
3. „ „ var. murrbyna, *Schw.*	f.	...	...	f.	...	r.
4. „ „ var. serrata, *Brady* ...	r.	...	v. r.	...	...	r.
5. „ tubulosa, *Costa*	f.	...	r.	c.	v. r.	f.
6. „ ringens (*Lam.*)	...	...	...	...	...	v. r.
7. „ „ var. striolata, *Brady*...	...	...	v. r.	v. r.	...	r.
8. „ comata, *Brady*	...	...	...	...	...	v. r.
9. Spiroloculina robusta, *Brady*	f.					
10. „ antillarum, *d'Orb.*	r.					
11. „ limbata, *d'Orb.*	v. r.					
12. „ grata, *Terquem*	v. r.	f.				
13. „ arenaria, *Brady*	...	...	...	r.		
14. „ asperula, *Karrer*	...	...	...	r.		
15. Miliolina trigonula (*Lam.*)	...	...	...	...	...	v. r.
16. „ insignis, *Brady*	...	v. r.				
17. „ tricarinata (*d'Orb.*)	f.	v. r.	...	...	r.	
18. „ circularis (*Bornemann*)	v. r.					
19. „ auberiana (*d'Orb.*) ·	r.	...	...	r.	v. r.	
20. „ cuvieriana (*d'Orb.*)	...	...	...	...	v. r.	v. r.
21. „ venusta (*Karrer*)	r.					

Species and Varieties.	I.	2.	3.	4.	5.	6.
22. Miliolina gracilis (d'Orb.)	v. r.					
23. „ amygdaloides, Brady	v. r.					
24. „ bicornis (W. & J.)	v. r.					
25. „ schreibersiana (d'Orb.)	...	v. r.				
26. „ undosa (Karrer)	f.					
27. „ linnæana (d'Orb.)	v. r.					
28. „ reticulata (d'Orb.)	f.					
29. „ parkeri, Brady	v. r.					
30. „ rupertiana, Brady	v. r.					
Subfamily HAUERININÆ.						
31. Ophthalmidium inconstans, Brady	v. r.					
32. Sigmoilina sigmoidea (Brady)	v. r.					
33. „ celata (Costa)	c.	...	r.	r.	r.	r.
Subfamily PENEROPLIDINÆ.						
34. Cornuspira carinata (Costa)	v. r.					
35. Orbitolites complanata, Lam.	c.					
36. „ marginalis (Lam.)	r.					
Subfamily ALVEOLININÆ.						
37. Alveolina melo (F. & M.)	v. r.					
38. „ boscii (Defr.)	c.	r.				
Family ASTRORHIZIDÆ.						
Subfamily PILULININÆ.						
39. Technitella melo, Norman	...	...	...	v. r.		
40. „ raphanus, Brady	...	...	...	f.		
41. Bathysiphon filiformis, M. Sars	...	...	...	v. r.		
Subfamily SACCAMMININÆ.						
42. Psammosphæra fusca, Schulze	v. r.					
43. Saccammina sphærica, M. Sars	r.					
44. „ socialis, Brady	..	...	...	...	v. r.	
Subfamily RHABDAMMININÆ.						
45. Hyperammina elongata, Brady	r.	...	...	c.	v. r.	r.
46. „ ramosa, Brady	f.	...	r.	v. c.	...	r.
47. „ arborescens (Norman)	r.	...	...	v. r.		
48. Marsipella elongata, Norman	...	...	...	r.		
49. Rhabdammina discreta, Brady	v. r.	...	...	f.	...	v. r.
50. Rhizammina indivisa, Brady	v. r.	...	...	c.	v. r.	
Family LITUOLIDÆ.						
Subfamily LITUOLINÆ.						
51. Reophax difflugiformis, Brady	v. r.	...	...	v. r.	v. r.	v. r.
52. „ scorpiurus, Montfort	v. r.	...	...	f.	...	f.
53. „ spiculifera, Brady	r.	...	...	v. r.	...	v. r.
54. „ distans, Brady	..	...	...	v. r.		
55. „ nodulosa, Brady	f.	...	...	...	...	v. r.
56. „ dentaliniformis, Brady	r.	...	v. r.	v. r.	v. r.	
57. „ bacillaris, Brady	v. r.					
58. „ pilulifera, Brady	c.					

Species and Varieties.	Samples.					
	1.	2.	3.	4.	5.	6.
59. Haplophragmium glomeratum, *Brady* ...	...	...	...	...	...	r.
60. „ latidorsatum (*Bornemann*) ...	r.	...	...	c.	...	v. r.
61. „ globigeriniforme (*P. & J.*) ...	r.	...	...	...	...	r.
62. „ canariense (*d'Orb.*) ...	v. r.	...	...	v. r.	...	r.
63. „ turbinatum, *Brady* ...	v. r.	...	...	v. r.	...	v. r.
64. „ rotulatum, *Brady* ...	v. r.	...	...	r.	...	v. r.
65. „ scitulum, *Brady* ...	r.	...	...	...	v. r.	r.
66. „ emaciatum, *Brady* ...	...	...	...	...	v. r.	
67. „ agglutinans (*d'Orb.*) ...	v. r.					
68. „ truncatuliniforme, sp. nov. ...	...	...	...	...	...	v. r.
69. Placopsilina cenomana, *d'Orb* ...	...	...	...	...	...	v. r.
Subfamily TROCHAMMININÆ.						
70. Thurammina papillata, *Brady* ...	...	...	...	r.		
71. Hormosina carpenteri, *Brady* ...	v. r.	...	...	c.	...	f.
72. „ ovicula, *Brady* ...	v. r.					
73. „ globulifera, *Brady* ...	...	...	...	...	...	v. r.
74. Ammodiscus incertus (*d'Orb.*) ...	r.	...	...	r.	...	v. r.
75. „ tenuis, *Brady* ...	...	...	...	...	...	v. r.
76. „ charoides (*J. & P.*) ...	v. r.					
77. Trochammina trulissata, *Brady* ...	f.	...	...	f.	...	r.
78. Webbina clavata, *J. & P.* ...	...	...	...	f.	...	v. r.
Subfamily LOFTUSINÆ.						
79. Cyclammina pusilla, *Brady* ...	r.	...	v. r.	f.	...	r.
80. „ cancellata, *Brady* ...	...	...	...	...	...	r.
Family TEXTULARIIDÆ.						
Subfamily TEXTULARIINÆ.						
81. Textularia sagittula, *Defrance* ...	...	...	...	r.	v. r.	v. r.
82. „ „ var. fistulosa, *Brady*.	r.					
83. „ gramen, *d'Orb.* ...	f.					
84. „ lythostrotum (*Schwager*) ...	f.	v. r.	...	r.	...	c.
85. „ conica, *d'Orb.* ...	...	...	...	...	...	v. r.
86. „ agglutinans, *d'Orb.* ...	r.	...	...	...	...	c.
87. Verneuilina pygmæa (*Egger*) ...	f.	...	...	r.		
88. „ propinqua, *Brady* ...	v. r.	...	...	r.		
89. Chrysalidina dimorpha, *Brady* ...	...	r.	...	...	v. r.	
90. Gaudryina pupoides, *d'Orb.* ...	r.	...	...	v. r.	...	v. r.
91. „ rugosa, *d'Orb.* ...	f.	...	r.	c.	...	f.
92. „ subrotundata, *Schwager* ...	c.	...	...	f.	...	c.
93. „ baccata, *Schwager* ...	...	v. r.				
94. „ siphonella, *Reuss* ...	...	...	...	...	v. r.	v. r.
95. Valvulina conica, *P. & J.* ...	r.	...	...	...	...	v. r.
96. Clavulina communis, *d'Orb.* ...	f.	...	f.	c.	r.	r.
97. „ parisiensis, *d'Orb.* ...	v. r.					
98. „ angularis, *d'Orb.* ...	r.					
Subfamily BULIMININÆ.						
99. Bulimina ovata, *d'Orb.* ...	f.	...	...	r.	v. r.	
100. „ pyrula, *d'Orb.* ...	f.	...	r.	...	v. r.	

Species and Varieties.	Samples.					
	1.	2.	3.	4.	5.	6.
101. Bulimina elongata, *d'Orb.*	v. r.					
102. „ pupoides, *d'Orb.*	v. r.	...	v. r.			
103. „ affinis, *d'Orb.*	r.					
104. „ elegans, *d'Orb.*	v. r.					
105. „ subcylindrica, *Brady*	r.					
106. „ declivis, *Reuss*	v. r.					
107. „ contraria (*Reuss*)	f.	...	...	r.	v. r.	f.
108. „ aculeata, *d'Orb.*	v. c.	...	r.	...	f.	v. r.
109. „ buchiana, *d'Orb.*	v. r.	...	...	v. r.	v. r.	
110. „ inflata, *Seguenza*	v. r.	...	...	...	...	v. r.
111. „ subornata, *Brady*	...	v. r.	...	...	v. r.	
112. „ rostrata, *Brady*	...	...	...	...	r.	
113. Virgulina schreibersiana, *Czjzek*	v. r.	...	...	...	v. r.	
114. „ subsquamosa, *Egger*	r.					
115. „ subdepressa, *Brady*	...	...	...	...	v. r.	
116. Bolivina punctata, *d'Orb.*	v. r.					
117. „ textilarioides, *Reuss*	...	...	...	...	v. r.	
118. „ limbata, *Brady*	v. r.	...	...	...	v. r.	
119. „ nobilis, *Hantken*	v. r.	...	...	..	r.	
120. „ beyrichi, *Reuss*	v. r.					
121. „ obsoleta, *Eley*	r.	...	...	..	v. r.	
122. „ robusta, *Brady*	...	...	...	...	v. r.	
123. „ arenosa. sp. nov.	v. r.					
124. Pleurostomella subnodosa, *Reuss*	r.					
125. „ alternans, *Schwager*	...	...	...	...	...	r.
Subfamily CASSIDULININÆ.						
126. Cassidulina murrhyna (*Schwager*)	...	...	v. r.	f.	v. r.	f.
127. „ calabra (*Seguenza*)	...	...	...	...	r.	c.
128. „ subglobosa, *Brady*	v. r.					
129. „ bradyi, *Norman*	r.					
130. „ parkeriana, *Brady*	...	...	...	...	...	v. r.
131. „ lævigata, *d'Orb.*	r.	...	...	...	v. r.	
132. Ehrenbergina serrata, *Reuss*	r.	f.	...	...	f.	
Family CHILOSTOMELLIDÆ.						
133. Chilostomella ovoidea, *Reuss*	r.	...	v. r.			
134. Allomorphina trigona, *Reuss*	...	...	...	...	...	v. r.
Family LAGENIDÆ.						
Subfamily LAGENINÆ.						
135. Lagena lævis (*Mont.*)	r.					
136. „ globosa (*Mont.*)	...	...	...	...	...	v. r.
137. „ apiculata, *Reuss.*	r.					
138. „ distoma. *P. & J.*	v. r.					
139. „ hispida, *Reuss*	v. r.					
140. „ aspera, *Reuss*, var. spinifera, nov.	...	...	...	v. r.		
141. „ sulcata (*W. & J.*)	...	...	...	...	...	v. r.
142. „ gracilis, *Will.*	v. r.	...	...	...	v. r.	
143. „ feildeniana, *Brady*	...	...	v. r.			
144. „ desmophora, *O. Ry. Jones*	r.	...	v. r.	...	v. r.	
145. „ hexagona (*Will.*)	r.					
146. „ marginata (*W. & J.*)	...	...	...	v. r.	r.	v. r.
147. „ „ var. catenulosa, nov.	...	...	...	...	...	v. r.

Species and Varieties.	1.	2.	3.	4.	5.	6.
148. Lagena seminiformis, *Schwager*					v. r.	
149. „ lagenoides (*Will.*)					v. r.	
150. „ capillosa (*Schwager*)						v. r.
151. „ fimbriata, *Brady*	v. r.					
152. „ castrensis, *Schwager*	v. r.			r.		
153. „ staphyllearia (*Schwager*)	v. r.					
154. „ alveolata, var. substriata, *Brady*	v. r.					v. r.
155. „ quadricostulata, *Reuss*	v. r.					v. r.
156. „ lævigata (*Reuss*)	v. r.					
157. „ orbignyana (*Seguenza*)	r.				r.	f.
158. „ formosa, *Schwager*	r.					v. r.
159. „ trigono-ornata, *Brady*	r.					
160. „ quadralata, *Brady*	v. r.					
161. Nodosaria (Dentalina) calomorpha, *Rss.*				r.		
162. „ radicula (*L.*)	r.					r.
163. „ pyrula, *d'Orb.*	v. r.				v. r.	
164. „ (D.) farcimen, *Reuss*	v. r.					
165. „ (D.) filiformis (*d'Orb.*)	v. r.					
166. „ (D.) roemeri (*Neugeboren*)	r.			v. r.		
167. „ (D.) communis, *d'Orb.*	c.		v. r.		r.	r.
168. „ (D.) consobrina (*d'Orb.*)	r.	v. r.				
169. „ (D.) inflexa, *Reuss*					r.	
170. „ ovulata, *Sherborn & Chapman*					r.	
171. „ (D.) soluta, *Reuss*				v. r.		r.
172. „ (D.) „ var. subaculeata, nov.	r.		v. r.	r.	v. r.	
173. „ (D.) acicula (*Lam.*)						v. r.
174. „ scalaris (*Batsch*)	v. r.					
175. „ „ var. separans, *Brady*	v. r.					
176. „ (D.) obliqua (*L.*)	v. r.					
177. „ raphanus (*L.*)	v. r.				r.	
178. „ (D.) adolphina (*d'Orb.*)	v. r.					v. r.
179. „ (D.) subcanaliculata (*Neug.*)					v. r.	
180. „ (D.) intercellularis (?), *Brady*					v. r.	
181. Rhabdogonium tricarinatum (*d'Orb.*)					v. r.	
182. Marginulina glabra, *d'Orb.*	v. r.				v. r.	
183. Cristellaria rotulata (*Lam.*)	v. r.	v. r.				v. r.
184. „ cultrata (*Montf.*)	f.	r.				v. r.
185. „ orbicularis (*d'Orb.*)			v. r.			
186. „ reniformis, *d'Orb.*						v. r.
187. „ tenuis (*Bornem.*)	v. r.					
188. „ obtusata, *Reuss*, var. subalata, *Brady*	r.					v. r.
189. „ crepidula (*F. & M.*)						v. r.
Subfamily POLYMORPHININÆ.						
190. Polymorphina angusta, *Egger*	v. r.					
191. „ ovata, *d'Orb.*				v. r.		v. r.
192. „ fusiformis (*Römer*)						v. r.
193. „ communis, *d'Orb.*	v. r.					
194. „ sororia, *Reuss* (fistulose var.)	v. r.					
195. Uvigerina interrupta, *Brady*	v. r.					
196. „ tenuistriata, *Reuss*	v. r.	r.			v. r.	
197. „ pygmæa, *d'Orb.*	r.				v. r.	

Species and Varieties.	Samples.					
	1.	2.	3.	4.	5.	6.
198. Urigerina aculeata, *d'Orb.*	c.	...	..	...	...	f.
199. „ angulosa, *Will.*	...	r.	...	...	r.	
200. „ „ var. spinipes, *Brady.*	...	v. r.				
201. „ asperula, *Czjzek*	...	v r.	f.	v. r.	r.	f.
202. „ „ var. ampullacea, *Brady*	...	...	...	...	r.	v. r.
203. „ schwageri, *Brady*	...	f.				
204. „ canariensis, *d'Orb.*	...	v. r.	...	...	r.	
205. „ brunnensis, *Karrer*	...	...	...	v. r.		
206. Sagrina columellaris, *Brady*	v. r.					
207. Ramulina globulifera, *Brady*	...	...	...	...	v. r.	
Family GLOBIGERINIDÆ.						
208. Globigerina bulloides, *d'Orb.*	c.	c.	...	f.	r.	c.
209. „ „ var. triloba, *Reuss*	...	v. r.	r.	...	v. r.	r.
210. „ dubia, *Egger*	...	...	r.	...	v. r.	v. r.
211. „ rubra, *d'Orb*	v. r.	r.	...	r.	...	r.
212. „ cretacea, *d'Orb*	c.	c.	v. r.	f.	r.	f.
213. „ conglobata, *Brady*	c.	c.	...	c.	r.	f.
214. „ æquilateralis, *Brady*	v. c.	r.	v. r.	c.	r.	c.
215. „ sacculifera, *Brady*	v. c.	r.	f.	c.	r.	r.
216. „ digitata, *Brady*	v. c.	...	...	v. c.	f.	v. r.
217. Orbulina universa, *d'Orb.*	v. c.	...	r.	c.	c.	v. c.
218. Hastigerina pelagica (*d'Orb.*)	...	...	...	...	...	v. r.
219. Pullenia obliquiloculata, *P. & J.*	v. c.	f.	r.	c.	c.	f.
220. „ sphæroides (*d'Orb.*)	...	...	...	...	r.	c.
221. „ quinqueloba, *Reuss*	...	...	...	...	...	v. r.
222. Sphæroidina bulloides, *d'Orb.*	v. r.	...	v. r.	r.	...	r.
223. „ dehiscens, *P. & J.*	c.	...	r.	f.	...	r.
224. Candeina nitida, *d'Orb.*	...	...	...	...	v. r.	
Family ROTALIIDÆ.						
Subfamily ROTALIINÆ.						
225. Cymbalopora poeyi (*d'Orb.*)	v. r.	r.				
226. „ (Tretomphalus) bulloides (*d'Orb.*)	v. r.					
227. Discorbina ventricosa, *Brady*	v. r.					
228. „ parisiensis (*d'Orb.*)	...	...	...	...	v. r.	
229. „ rosacea (*d'Orb.*)	...	...	...	...	v. r.	v. r.
230. „ rugosa (*d'Orb.*)	...	...	...	...	...	r.
231. Planorbulina acervalis, *Brady*	v. r.	v. r.				
232. „ larvata, *P. & J.*	...	...	...	...	...	v. r.
233. Truncatulina lobatula (*W. & J.*)	r.	...	...	...	v. r.	r.
234. „ wuellerstorfi (*Schwager*)	v. c.	...	c.	c.	r.	r.
235. „ pygmæa, *Hantken*	r.	...	...	...	...	r.
236. „ ungeriana (*d'Orb.*)	f.	r.	r.	c.	r.	f.
237. „ haidingerii (*d'Orb.*)	r.	...	...	...	r.	v. r.
238. „ robertsoniana, *Brady*	r.	...	...	...	...	v. r.
239. „ dutemplei (*d'Orb.*)	r.					
240. „ præcincta (*Karrer*)	...	r.				
241. Truncatulina akneriana (*d'Orb.*)	...	...	v. r.			
242. „ tenera, *Brady*	...	...	...	r.	v. r.	v. r.
243. „ culter (*P. & J.*)	...	...	...	...	...	v. r.
244. Anomalina grosserugosa (*Gümbel*)	r.	...	...	...	...	f.